BRITISH AND OTHER FRESHWATER CILIATED PROTOZOA

A NEW SERIES
Synopses of the British Fauna
No. 22

Edited by Doris M. Kermack and R. S. K. Barnes

BRITISH AND OTHER FRESHWATER CILIATED PROTOZOA

Part I

Ciliophora: Kinetofragminophora

Keys and notes for the identification of the free-living genera

COLIN R. CURDS

Department of Zoology, British Museum (Natural History), London SW7 5BD

1982
Published for
The Linnean Society of London
and
The Estuarine and Brackish-water Sciences Association
by
Cambridge University Press
Cambridge
London   New York   New Rochelle
Melbourne   Sydney

Published by the Press Syndicate of the University of Cambridge
The Pitt Building, Trumpington Street, Cambridge CB2 1RP
32 East 57th Street, New York, NY 10022, USA
296 Beaconsfield Parade, Middle Park, Melbourne 3206, Australia

First published 1982

Printed in Great Britain at the Pitman Press, Bath

British Library Cataloguing in Publication Data
Curds, Colin R.
British and other freshwater ciliated protozoa.
Pt. 1: Ciliophora: Kinetofragminophora. – (Synopses of the British Fauna. New
series; 22)
1. Ciliata–Identification 2. Ciliata–Great Britain–Identification
I. Title II. Linnean Society of London III. Estuarine and Brackish-water Sciences
Association IV. Series
593.17′2′0941 QL368.C5
ISBN 0 521 24257 6 hard covers ✓
ISBN 0 521 28558 5 paperback

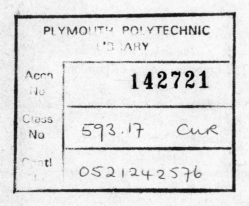

Foreword

The two *Synopses* devoted to *British and other freshwater ciliated Protozoa* (Nos. 22 & 23) have been published exactly a century after the appearance between 1880–2 of W. Saville Kent's classic *Manual of the Infusoria*. Dr Curds is a member of the scientific staff of the British Museum (Natural History) as was Saville Kent when he wrote his manual, emphasising the important role of this great national museum in the field of taxonomy and the importance that it has placed upon the collection of living material as well as upon the maintenance of its vast collections of dead specimens.

The remarkable resistance of ciliated protozoa to desiccation by cyst formation has led to their easy transport in the air and on the outsides of other animals, hence their lack of respect for national boundaries. Thus this *Synopsis* is not restricted to the British kinetofragminophorans but includes those from elsewhere. It is however restricted to their genera: to run down these protozoa to species often involves elaborate techniques with expensive apparatus outside the scope of this *Synopses* series, which is designed to be a set of handbooks for use in the field and laboratory by amateur and professional biologists from VIth-form level upwards. Extending the coverage to specific level would also increase their already generous size so they could no longer be described as 'handbooks'. The reference list at the back of each *Synopsis* should be of assistance if a particular specimen needs to be 'tracked down' to species level.

Although there are two *Synopses* devoted to freshwater ciliates, each is complete in itself; the introductory chapters and the associated figures (pp. 1–59) will be repeated with minor changes in each part.

The editors acknowledge, as has the author, their gratitude to the Natural Environment Research Council for a grant to cover the artist's fees. They also thank both author and artist for the skill and care with which they have used their respective skills to produce a work which is both scientifically accurate as well as aesthetically pleasant to read and use.

R. S. K. Barnes

Estuarine and Brackish-water
Sciences Association

Doris M. Kermack

The Linnean Society

A Synopsis of the British Freshwater Ciliates Part I. The Kinetofragminophora

COLIN R. CURDS

Department of Zoology, British Museum (Natural History), London SW7 5BD

Contents

Introduction

Ciliated protozoa are found in all moist habitats. They are generally cosmopolitan in their distribution and may be free-living or symbionts. They are extremely common, frequently numerous and indeed it is rare to find a sample of natural water without some ciliates being present. In spite of their abundance they are recognised to be a difficult group to identify and this is partly due to the lack of a suitable key both for specialist and non-specialist alike. The now classical keys of Kahl (1930–35, 1934) are still used by most specialists for identification purposes; while these have the merit of being keys to species they are long outdated. Patterson (1978) has recently published an English translation of the keys to the genera in one of Kahl's (1930–35) works. The non-specialist has tended to use Kudo (1966) which while containing many generic descriptions does not include them all nor does it contain keys and is also rather out of date.

Until recently ciliates were classified as a subphylum within the phylum Protozoa. It is now generally recognised that ciliates are so distinctive and diverse as to merit the rank of phylum in their own right. There are about 1125 described genera comprising some 7200 species. With so many taxa now available it is not easy for a second 'Kahl' to be written, indeed it would need a team of specialists writing within their own small fields to accomplish the task within a reasonable length of time. Here to keep the number of taxa down to a manageable figure it was necessary to include only freshwater free-living genera as listed in Corliss (1979). Thus all freshwater ciliates living freely or growing as symphoriants on the outside of aquatic animals and plants have been dealt with but no exclusively entozooic genera have been included. Similarly no ciliate genera reported from marine or brackish waters have been described herein unless at least one species has been identified in freshwater. We realise that these restrictions are partly arbitrary but even so ≃400 genera have been included in the two *Synopses* devoted to freshwater ciliates.

Although these form part of the series of *Synopses of the British Fauna*, a British ciliate is not easy to define since ciliates are so cosmopolitan and this is complicated by so few protozoologists having identified ciliates within the British Isles. We have therefore included *all* free-living freshwater genera as all are at least potentially to be found in British waters. The only likely exceptions to this are certain ciliates which grow as symphoriants on specific species of aquatic animals. For example, there are several chonotrichs and suctoria which have been found growing on the exoskeleton of specific crustaceans endemic to Lake Baikal in Russia. Since the numbers of these

1

examples are so low it seemed worthwhile retaining them since the key may then be used on a world-wide basis.

We have aimed the key at both the specialist and non-specialist. Ciliate taxonomy has moved so rapidly in the last decade that it is difficult for the ecologist to keep abreast of these changes. It has been our intention to provide an up-to-date résumé. We have purposely not included the specific names of ciliates in diagrams and descriptions unless necessary to do so for some nomenclatorial reason, this will prevent the novice from recording species from any examples given, which evidently was the case in the past with many non-specialists using works such as Kudo (1966). However, we do wherever possible cite references to descriptions of species and keys to species where they exist. Similarly because size varies considerably from species to species within a genus we have not included scales on diagrams but have indicated in the text when the size of the species within a particular genus is particularly large or small.

Since much of the skill in identification lies in knowing precisely what features to look for, we have included an extensive introductory review on the morphology and biology of these animals. Similarly a review on staining and handling techniques is included since at some stage during an identification it is often necessary to display a particular structure by staining.

The original drawings of Figs. 41–243 are deposited in the Protozoa Section, Department of Zoology, British Museum (Natural History).

General structure and biology

All ciliates are single-celled microscopic animals, although their overall range is about 15–2000 μm long most species lie within the range of 40–200 μm long. The shape of the **body** is generally simple, often ovoid or spherical but there are many variations and the terms most commonly used in the descriptions given in this book are defined in Fig. 1. It should be pointed out that it is rare for a ciliate to be completely and regularly spherical or ovoid, they may be slightly elongated or compressed and thus the terms used in the descriptions are the nearest equivalents. Many genera are rounded in cross-section but others may be flattened either dorso-ventrally or laterally and the position of the **oral opening** is usually used to define the ventral surface. Thus where the flattening is lateral the oral aperture will lie on an edge rather than on a surface and in this case there are ventral and dorsal edges and lateral surfaces. In certain flattened ciliates the opening may lie so closely to an edge that it is difficult to know if it is on a surface or along an edge; in cases like these the position of the **stomatogenic kinety** is often considered to be more important than the position of the oral aperture for defining the ventral aspect of the cell. The longitudinal axis is defined by the direction of the kineties which generally run along the longitudinal body axis although there is frequently some spiralling present. The **kinetodesmata** always lie on the animal's right of the line of kinetosomes so that the anterior and posterior can be defined (Fig. 2A). In practice the anterior end is usually held forwards whilst swimming. Perhaps it should be noted here that right and left in descriptions throughout the text refer to those sides of the animal not of the observer.

The outside of all ciliates is covered by a cell wall or **pellicle**, many remain highly plastic but in other cases the pellicle is thicker and/or made more rigid by the addition of **ribs** or **ridges** and this is often referred to as body armour or the **cuirass**. Similarly, several genera with armour possess spines or spikes which frequently project from the body. Other ciliates retain their plasticity, but produce a shell-like structure known as a **lorica** inside which the animal is protected. Some ciliates live permanently trapped inside the lorica, others leave sporadically to return later. Several ciliate genera attach themselves to the base of the lorica and these may then extend out of the lorica to feed and contract back inside for protection. Loricas may be membranous, gelatinous or pseudochitinous and they commonly include debris from the local environment in their construction. Many ciliate genera produce **cysts** for protective purposes, usually to withstand desiccation or some other environmental hazard. Protective cysts have thick walls but there are also certain

3

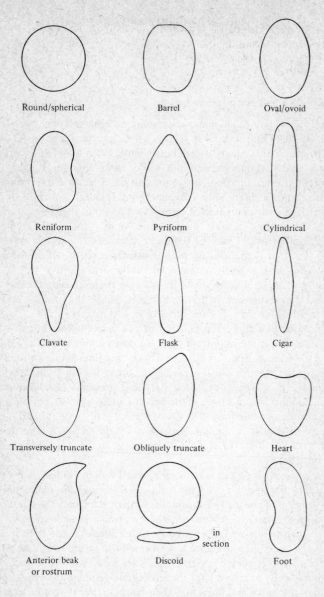

Fig. 1. Shapes of ciliates. Organisms will rarely be exact replicas, the nearest approximation will be given in the text. When two descriptive names are used in the above diagram they represent the two- and three-dimensional equivalents respectively.

genera that can produce thin-walled reproductive cysts. The best known examples of these include the genus *Colpoda* and its relatives which undergo several divisions within the cyst to produce up to eight daughter cells which then emerge.

Some ciliates are able to protect themselves by means other than the possession of a solid external covering; certain genera secrete a mucilage from subpellicular vesicles called **mucocysts** and others defend themselves by the expulsion of fibrous harpoon-like **trichocysts**. These two organelles are not easily distinguished by light microscopy and for this reason we have used the all-embracing term trichocyst throughout the key. True fibrous explosive trichocysts are restricted to the peniculine hymenostomes and the hypostome nassulid suborder Microthoracina.

The majority of ciliates swim freely in the water but there are many others which are attached to surfaces of some kind, these surfaces may be inanimate objects or other living organisms such as plants and animals. Attachment to the surface may be direct or via a stalk or the lorica. **Stalks** may be contractile or non-contractile, and they may be branched giving rise to colonial forms or they may be unbranched.

Tentacles and tentacle-like structures appear in several subclasses. In the suctoria they represent the sites for ingestion, there being no other oral aperture and no cilia present. However, in the hypostomes, gymnostomes and vestibulifera there are always cilia present, the tentacle-like structures do not represent the ingestion site and there is always a single oral aperture situated elsewhere on the body.

All ciliates, with very few exceptions, possess **cilia** at some stage in their life history. Cilia are fine hair-like organelles (about 0.25 μm diameter, length variable but often 7–10 μm long) that project out from the body from a subpellicular basal granule known as a **kinetosome** (Fig. 2A). Cilia beat rhythmically for locomotion and to create feeding currents. In the rare cases where cilia are absent either completely or, more commonly, during a particular stage in the life cycle their kinetosomes always persist. Kinetosomes are usually arranged in rows along the body axis, and a single row of them is known as a **kinety** (Fig. 2A). There are usually several kineties arranged on the body surface and associated with these structures are longitudinally orientated subpellicular cytoplasmic fibrils known as **kinetodesmata** (Fig. 2A) which arise close to the base of a kinetosome and extend anteriorly on the right of the kinety concerned. Individual kinetodesmal fibrils are relatively short, thus the longitudinal fibre travelling along the right-hand side of a kinety is not a continuous fibre as is apparent from the light microscope, but a series of over-lapping fibrillar units. The complete collection of kineties plus any oral ciliary organelles that may be present are referred to collectively as the **infraciliature** which may be displayed by the application of various silver-impregnation techniques (see *Practical methods*) and is therefore frequently referred to as the 'silver-line system'. In certain ciliates, particularly the spirotrichs, cilia may be reduced and largely replaced by **cirri** (Fig. 3C). These latter organelles are complexes of numerous long cilia loosely grouped together to form stout, tapering organelles that are rounded in cross-section. Although coordinated, cirri do not beat uniformly but are often used for 'walking' over solid surfaces.

With the exception of the suctoria* and the astomes (only one free-living representative) all ciliates possess a single **oral aperture** (Fig. 2A,B,C). We have purposely used the term oral aperture to denote the actual hole, slit or indentation in the body without any regard to any oral ciliary structures which may be present since the term loosely includes all of the precise protozoological terms of **cytostome, buccal overture** and **vestibular aperture**. While the technical differences between these structures can be precisely defined (see below) we would point out that it is not usually easily possible for the microscopist to be able to differentiate between a buccal cavity and a vestibulum without considerable effort. For example, it would be simple to key out the Vestibulifera by the presence of a vestibulum, however, in practice this would be valueless to the user since it is not easy to recognise a vestibulum. However, the position and nature of the oral aperture can also be of great importance for identification. In most cases the oral aperture lies somewhere in the anterior body half but there are several examples of subequatorial oral apertures. When the position of an oral aperture is of importance in the key, several illustrative examples are usually given.

The oral region of many ciliates contains compound ciliary organelles that are specialised for feeding although some in the course of time have regained their locomotory function. These compound structures are of two types and are called **undulating membranes** and **membranelles** according to their structure. An undulating membrane is simply a line or arc of cilia set close together, in a single row, so that they more or less permanently coalesce into a membrane. They are set on the right of the oral area. Conversely, membranelles are composed of two or three rows of cilia forming a block, the free ends of which adhere together to form triangular or trapezoidal flaps. They are typically arranged on the left of the oral area (Fig. 2A,C).

In the subclass Gymnostomata, most species are without oral ciliature and the **cytostome** is at, or near to, the surface of the body and located apically, subapically or laterally (Fig. 2D–F). The cytostome is the two-dimensional hole which marks the cell mouth proper denoting the end of any ciliation. Any passage beyond the cytostome is known as a **cytopharynx** and is always unciliated. The cytopharynx is frequently strengthened by a series of rods or **trichites** and the complex formed is known as the **cytopharyngeal apparatus** (Fig. 2D,F). The **trichites** or rods encircle the cytopharynx to form a basket-like structure which may be straight (**rhabdos** type, Corliss, 1979) or

---

* This branch of zoology has yet to develop a colloquial language and such words as 'suctoria' can be used in two contexts: firstly formally Suctoria, referring to the subclass of that name and secondly informally, when the lower-case or small 's' is used, suctoria. Protozoa and protozoa is another example.

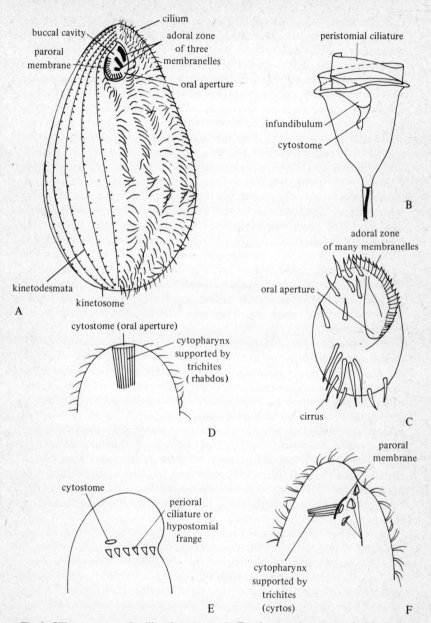

Fig. 2. Ciliary structures in ciliated protozoa. A, *Tetrahymena*, impregnated with silver on left, as seen in living specimen on right. B, *Vorticella*, a peritrich. C, *Euplotes*, a hypotrich. D, *Holophrya*, a simple gymnostome. E, *Nassula*, a hypostome. F, *Furgasonia*, a hypostome.

curved (**cyrtos** type). Although there are other morphological differences between the two types (see Corliss, 1979) these may only be distinguished by the use of an electron microscope. The straight rhabdos type is considered to be more primitive and is found in the gymnostome and vestibuliferan ciliates.

In the subclass Vestibulifera the cytostome is at the base of a depression in the body containing a ciliature which is more or less complex in appearance but which is predominantly somatic in nature and origin. The cytostome leads to a cytopharyngeal apparatus of the rhabdos type. The oral area in the subclass Hypostomata has, with the major exception of the chonotrichs, moved to the ventral surface and the cytostome may be at the base of a shallow depression known as an **atrium** which may be the forerunner of the buccal cavity which is found in the hymenostome ciliates (see below). The cytostome usually leads to a cytopharyngeal apparatus which is of the cyrtos type. In some, the ciliature around the oral aperture (the **perioral ciliature**) is in the form of an extensive band which is commonly referred to as the **hypostomial frange** (Fig. 2E). This frange may be considerably reduced in some forms to three short **pseudomembranelles** (Fig. 2F) which are located on the left of the oral aperture and may lie within an atrium. In certain species there is even a fourth '**paroral membrane**' (Fig. 2F) on the right such that the oral ciliature as a whole at least superficially looks very much like the undulating membrane and three membranelles that are found in the hymenostomes.

Species in the subclass Hymenostomata stand neatly between the 'lower' ciliates above and those that are to be treated below. While there is always a definite oral ciliature, it is usually inconspicuous and composed of only three or four specialised membranes or membranelles which are located in a **buccal cavity**. Commonly there is an **adoral zone of three membranelles** (Fig. 2A) on the left and a single **paroral membrane** on the right. In others there are **peniculi** deep within the buccal cavity, these are compound ciliary organelles in the form of a long band of often short cilia typically found on the left wall of the buccal cavity. Peniculi are responsible for creating the strong feeding currents seen in ciliates such as *Paramecium*. In other hymenostomes, the paroral membrane (sometimes in multiple segments) on the right of the buccal area becomes dominant and the buccal cavity may be distinct or shallow and difficult to distinguish.

Members of the subclass Peritricha are highly specialised and are immediately recognisable. Here the oral field covers the entire apical end of the body and the **peristomial ciliature** (Fig. 2B), which winds anticlockwise and plunges deep into an **infundibulum** towards the cytostome. The final group of ciliates are the spirotrichs and their oral ciliature is dominated by a well developed conspicuous **adoral zone of many membranelles (AZM)** (Fig. 2C) which often extend out onto the surface of the body. These membranelles are used both for feeding and locomotion.

Ciliates feed on a variety of food sources although it is generally in the form of other microorganisms such as bacteria, algae and other protozoa. In

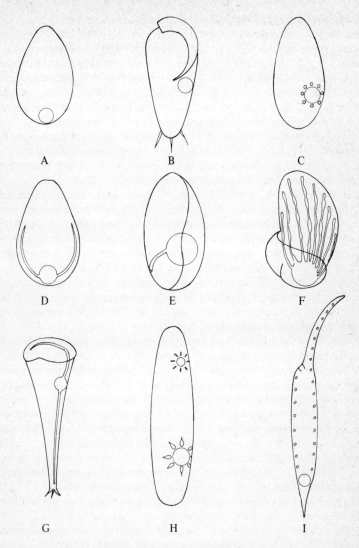

Fig. 3. Contractile vacuoles in ciliates, their location and their accessory structures. A, *Tetrahymena*, single terminal vacuole. B, *Stylonychia*, single lateral vacuole. C, *Bursostoma*, vacuole with satellite vesicles. D, *Climacostomum*, vacuole with two serving canals. E, *Lembadion*, vacuole with expulsion canal. F, *Tillina*, vacuole with several serving canals. G, *Stentor*, vacuole with two long serving canals. H, *Paramecium*, two vacuoles with radiating serving canals. I, *Dileptus*, large terminal and many lateral vacuoles.

rarer cases, other invertebrates such as flatworms may be ingested. In all cases with the exception of the suctoria the food is collected by means of a ciliary current and it is gathered at the cytostome where it is engulfed to form a food vacuole. In the case of suctoria, the food organism is usually another ciliate which is sucked down through the hollow tentacles into the predator's body. **Food vacuoles** are usually spherical although lemon-shaped vacuoles are frequently seen in the peritrichs. Food is digested within the vacuoles which move around the cell following a more or less distinct pathway before undigested remains are voided to the exterior via a permanent pore or **cytoproct**. It should be noted that when a ciliate is feeding upon green algae the latter should not be confused with mutualistic zoochlorellae which are found in certain ciliate species. When algae form the food source the green areas will be present in discrete spherical clumps whereas mutualistic algae will be distributed generally throughout the cytoplasm and not in clumps.

Excess water is expelled from ciliates by the pulsating action of transparent vacuoles known as **contractile vacuoles**. Water is discharged to the exterior via a pore, or pores, which may in some species open into the contractile vacuole via a canal. Contractile vacuoles may be solitary (Fig. 3A–G) or numerous (Fig. 3H–I), and the position, structure and number is often of great use for identification purposes. The most commonly encountered type is the solitary terminal (Fig. 3A) or lateral vacuole (Fig. 3B). Several genera have vacuoles with serving canals (Fig. 3D–G) or vesicles (Fig. 3C).

Ciliates have two kinds of nucleus. The much larger **macronucleus** is the vegetative or trophic nucleus of ciliates and it is concerned with the 'day-to-day running' of the cell. It controls the organism's phenotype. In many cases it is a single spherical or oval body that often can be displayed in the living cells by the use of phase-contrast microscopy although various staining techniques are more reliable (see *Practical methods*). The macronucleus may be multiple and sometimes diverse in shape (Fig. 4). The shape, number and location of macronuclei can often be of value for identification purposes. The macronucleus divides during asexual fission and is derived from the micronucleus during sexual reproduction. The **micronucleus** is very much smaller than the macronucleus and is invariably spherical; it is the generative nucleus of ciliates and is concerned with sexual processes. There may be more than one micronucleus present. Although a ciliate can often survive indefinitely without a micronucleus, and there are several amicronucleate strains of *Tetrahymena* that have been growing for several decades, the presence of a macronucleus is vital to the ciliate's survival.

Most ciliates are able to reproduce both asexually and sexually. Asexual reproduction is the normal method of increasing the numbers of a species within a population and sexual processes serve to ensure the transference of genetic material from one organism to another.

Asexual reproduction in ciliates is usually by **binary fission**; this takes place transversely across the somatic kineties. However, other specialised methods of asexual reproduction, such as **budding**, are known to occur in certain

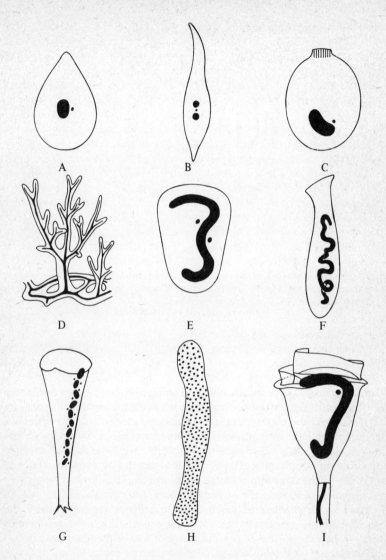

Fig. 4. Macro- and micronuclear diversity in ciliates. A, *Tetrahymena*. B, *Litonotus*. C, *Belonophrya*. D, *Dendrosoma* (only macronucleus shown). E, *Euplotes*. F, *Spathidium* (only macronucleus shown). G, *Stentor*. H, *Myriokaryon* (only macronuclei shown, up to 3000 per cell). I, *Vorticella*.

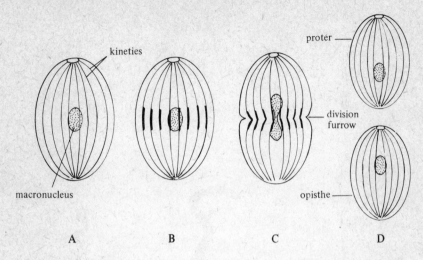

Fig. 5. Transverse binary fission in a primitive ciliate with telokinetal stomatogenesis. A, Ciliate prior to division. B, Ciliate at onset of division showing increased numbers of kinetosomes in equatorial region. C, Ciliate and macronucleus dividing. D, Daughter cells (proter and opisthe) immediately after division.

groups and these will be discussed later. During **transverse binary fission**, the micronucleus divides by mitosis and splits while the macronucleus undergoes DNA synthesis and constricts into two parts amitotically. Each daughter cell thus contains part of both micro- and macronucleus on division. The presence of cilia and other complex organelles in ciliates has far-reaching consequences on division. These structures are partly formed anew and partly derived by the transformation of the existing structures of the mother cell. Renewal of the **somatic** or **body ciliature**, for example, begins with an increase in the number of kinetosomes usually in the area where the transverse division, furrow or waist will later appear. The kinetosomes in this zone will give rise to the cilia of the posterior half of the anterior daughter cell (**proter**) as well as to the anterior ciliature of the posterior daughter (**opisthe**). The formation of the oral ciliature (**stomatogenesis**) varies considerably from group to group but four major methods are commonly recognised.

In the more primitive ciliates (Fig. 5) where there is little or no specialised oral ciliature, either all or some of the kinetosomes of the somatic kineties at the apex and equator are involved. Those in the equator which are transected at fission simply turn in at the anterior end of the forming opisthe to produce the new oral area. This type of stomatogenesis is called **telokinetal**. In the more advanced groups a single somatic kinety or several somatic kineties in

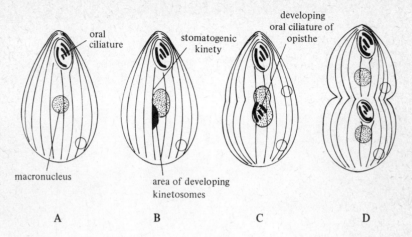

Fig. 6. Transverse binary fission in *Tetrahymena* a ciliate with parakinetal stomatogenesis. A, Ciliate prior to division. B, Ciliate at onset of division showing area of developing kinetosomes to left of stomatogenic kinety. C, Oral ciliature of opisthe developing from kinetosomal field. D, Ciliate just before final fission.

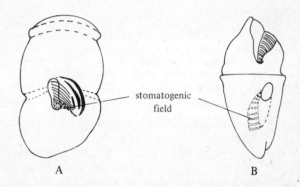

Fig. 7. Stomatogenesis in ciliates. A, *Urocentrum*, a ciliate with buccokinetal stomatogenesis. B, *Strombidium*, a ciliate with apokinetal stomatogenesis.

the central body region are involved (**stomatogenic kineties**). Here an apparently randomly organised field of cilia-free kinetosomes develop alongside or from the stomatogenic kineties (Fig. 6B) which then become properly organised (Fig. 6C) to form the oral ciliature of the opisthe. The oral ciliature of the mother cell may be completely or partially replaced or restructured simultaneously at the anteriormost ends of the stomatogenic kineties. This type is known as **parakinetal stomatogenesis** and is characteristic of some hymenostomes and some spirotrichs. In another type, **buccokinetal**

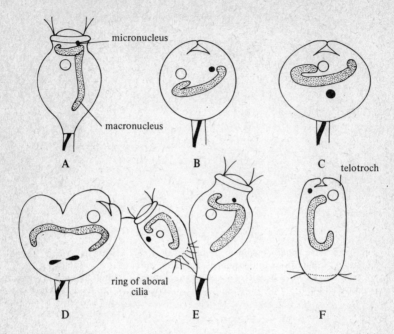

Fig. 8. Asexual binary fission in the peritrich *Vorticella*. A, Adult cell prior to division. B, Peristome closes over. C, Macro- and micronuclei begin to divide. D, Micronucleus divides, macronucleus dividing. E, Daughter-cell formed. F, Daughter-cell ciliated aborally, swims away as telotroch.

stomatogenesis, the fields of kinetosomes involved in the development of the oral ciliature of the opisthe are derived at least partly from the kinetosomes of the oral ciliature of the mother cell (Fig. 7A). This type is found in many hymenostomes, peritrichs, and in some spirotrichs. In the final major type of stomatogenesis (**apokinetal**) the field of kinetosomes suddenly appears without any apparent parental somatic or oral ciliature origins (Fig. 7B), this type is found in some spirotrichs.

The method of fission in peritrichs (Fig. 8) is exceptional in that it is not transverse but apparently longitudinal. In the solitary genera, the division is also unequal such that the daughter which swims away is smaller than the stationary mother cell. However, in the colonial peritrichs the division is equal and both daughters remain attached to the stalk of the mother colony. It seems likely that these variations are adaptations to their sedentary life-style.

Budding is another type of asexual division which is found in all the suctoria (Fig. 9) and also in the chonotrichs (Fig. 10). The adult chonotrich only has oral ciliature and the adult suctorian does not possess cilia at all. In

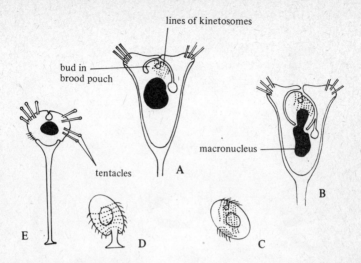

Fig. 9. Endogenous budding in the suctorian *Acineta*. A, Adult with bud in brood pouch, note field of kinetosomes. B, Part of macronucleus of adult passes into bud. C, Ciliated embryo escapes from brood pouch. D, Newly settled embryo secretes stalk. E, Embryo grows, loses cilia and gains tentacles.

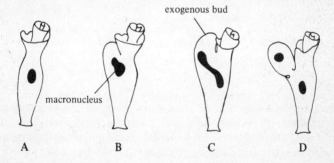

Fig. 10. Exogenous budding in the chonotrich *Spirochona*. A, Adult prior to budding. B,C, Macronucleus dividing, exogenous bud appearing. D, Macronucleus divided, ciliated embryo ready to leave parent.

both cases however, the daughter cells or larval forms are ciliated and their cilia are derived from kinetosomes which migrate from the mother cell. Buds may be produced externally (**exogenous buds**) as in all chonotrichs and some suctoria, or internally (**endogenous buds**) as in other suctoria. In addition there are several suctorian variations and these will be mentioned in the introduction to that group (p. 299).

Conjugation is the usual method of sexual reproduction in ciliated protozoa. Mating can occur only when two **mating types** (or sexes) of the same

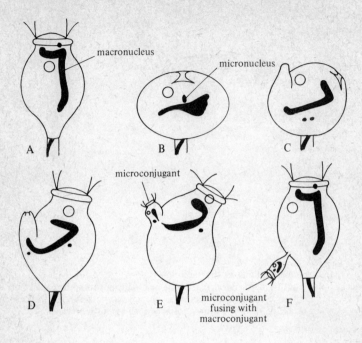

Fig. 11. Sexual reproduction in the peritrich *Vorticella*. A, Adult prior to conjugation. B, Macronuclear division. C, D, E, Microconjugant being formed by unequal division. F, Microconjugant swims and fuses completely with the macroconjugant.

species are mixed together. In most cases the sexes are morphologically indistinguishable from each other but there are some exceptions to this general rule. In the peritrichs, for example, a small **microconjugant** is budded from the stationary male cell from which it breaks free and swims to the female (**macroconjugant**) with which it completely fuses (Fig. 11). In solitary peritrich genera, such as *Vorticella*, the microconjugant is produced by a single unequal division but in most colonial genera, such as in *Epistylis*, several (2, 4 or 8) microconjugants are produced by multiple equal divisions. Furthermore, in certain suctoria where the conjugants may or may not be of equal size, only one, the smaller where there is a size difference, breaks free from the stalk after fusion and donates its nuclear material to the other conjugant which remains sessile.

Apart from the exceptions such as those outlined above, two similar conjugants come together and unite, usually along their oral surfaces, and remain joined until conjugation is complete. The best known example of conjugation is in *Paramecium aurelia* (Fig. 12). Here the macronucleus

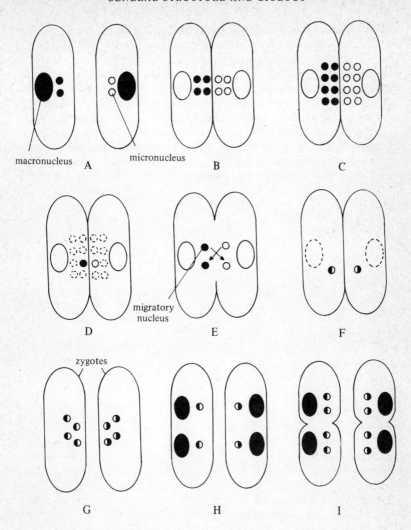

Fig. 12. Conjugation in *Paramecium aurelia* (see text for detail).

begins to break down and eventually disappears completely (Fig. 12F). In *P. aurelia* there are two micronuclei and these both divide twice by meiosis (Fig. 12B,C) to produce eight haploid micronuclei of which seven disintegrate (Fig. 12D). The remaining micronucleus in each conjugant divides by mitosis to produce two micronuclei in each cell. One of each pair (the **migratory nucleus**) migrates to the other cell (Fig. 12E) where it fuses with the **stationary nucleus**. After fusion (Fig. 12F), the two cells separate and the nucleus in each zygote divides twice mitotically (Fig. 12G); two of the four products become macronuclei, the other two remain as diploid micronuclei (Fig. 12H). At the next asexual binary fission, one macronucleus passes to each daughter cell while the micronuclei each divide mitotically so that the daughters have the original nuclear complement (Fig. 12I).

There are two other sexual processes found in ciliates which are types of self-fertilisation serving to reorganise the micronucleus. In **cytogamy**, the cells pair and their nuclei undergo the transformations as in conjugation but the cell wall between the conjugant does not break down and the migratory nucleus has to recombine with the stationary nucleus in its own cell. **Autogamy** is similar to cytogamy except that the cells do not pair at all.

Classification of ciliated protozoa

Ciliates are single-celled animals which until recently were classified as a subphylum within the phylum Protozoa. Of all the groups of protozoa the ciliates are the most clearly defined since they almost exclusively all possess the following five characteristics in common.

1. All, with very few exceptions, possess cilia at some stage during their life cycle. Cilia are fine hair-like organelles that beat rhythmically for locomotion and to produce feeding currents.
2. All, without exception, possess subpellicular structures known as kineto-somes from which the cilia arise. In rare cases when cilia are absent either completely or more commonly during a particular stage in the life cycle, their basal bodies or kinetosomes always persist.
3. During asexual fission the animal divides into two identical daughters. The fission line takes place at right angles across the lines of kinetosomes and is known as transverse binary fusion.
4. There are two types of nucleus, the larger macronucleus which is vegetative and is concerned with the 'day-to-day running' of the cell and the much smaller micronucleus which is involved in sexual processes.
5. Sexual exchange occurs during conjugation and autogamy.

Several classifications of the phylum are currently available (see review by Corliss, 1975), the one adopted here is that published most recently by Corliss (1977, 1979). The phylum is divided into three classes all of which contain both freshwater and marine species. Class 1, the Kinetofragminophora is included in this *Synopsis*, classes 2 and 3, the Oligohymenophora and Polyhymenophora are dealt with in *Synopsis No. 23*, but below is listed an outline of the classification used throughout both parts. It should be noted that the classification given only includes freshwater free-living ciliates so that certain exclusively marine or parasitic groups have been omitted.

Phylum CILIOPHORA

Class 1. KINETOFRAGMINOPHORA

Subclass 1. GYMNOSTOMATA
Order Karyorelectida
Order Prostomatida
Suborder Prostomatina
Suborder Prorodontina
Order Haptorida
Order Pleurostomatida

Subclass 2. VESTIBULIFERA
Order Trichostomatida
Order Colpodida
Order Bursariomorphida

Subclass 3. HYPOSTOMATA
Order Synhymeniida
Order Nassulida
Suborder Nassulina
Suborder Microthoracina
Order Cyrtophorida
Suborder Chlamydodontina
Suborder Dysteriina
Order Chonotrichida
Suborder Exogemmina
Order Apostomatida
Suborder Apostomatina

Subclass 4. SUCTORIA
Order Suctorida
Suborder Exogenina
Suborder Endogenina
Suborder Evaginogenina

Class 2. OLIGOHYMENOPHORA

Subclass 1. HYMENOSTOMATA
    Order Hymenostomatida
        Suborder Tetrahymenina
        Suborder Ophryoglenina
        Suborder Peniculina
    Order Scuticociliatida
        Suborder Philasterina
        Suborder Pleuronematina
    Order Astomatida

Subclass 2. PERITRICHA
    Order Peritrichida
        Suborder Sessilina
        Suborder Mobilina

Class 3. POLYHYMENOPHORA

Subclass SPIROTRICHA
    Order Heterotrichida
        Suborder Heterotrichina
        Suborder Armophorina
        Suborder Coliphorina
    Order Odontostomatida
    Order Oligotrichida
        Suborder Oligotrichina
        Suborder Tintinnina
    Order Hypotrichida
        Suborder Stichotrichina
        Suborder Sporadotrichina

Practical methods

Samples of water from lakes, ponds, streams, rivers, bogs, mosses, sewage-treatment processes, holes in trees and other places may be collected in any convenient glass or plastic container. Simply dipping the container into the water will not usually yield much material, so the active collection of submerged plant material, organic debris, bottom deposits, dead leaves and twigs, and surface scum from the surrounding water body is highly recommended. A Pasteur pipette with a large bore is often helpful in collecting bottom deposits. The larger planktonic forms may be collected by means of a net of the finest grade bolting silk although even this will not retain the smaller forms, thus larger volumes of water may have to be centrifuged or filtered back in the laboratory if these are to be studied. Many ciliates, particularly sedentary or crawling forms, may be captured on glass slides, cover slips or Petri dishes by submerging them in the natural aquatic environment or aquarium for a week or more. A recent novel variation of this latter method is the use of submerged man-made domestic sponges; after a few weeks the protozoa in the interstices of the sponge may be recovered by gentle squeezing into an appropriate container. The collection of various other invertebrate animals should not be forgotten as there are several ciliate genera only to be found growing epizooically on animals such as crustaceans, molluscs etc.

It is imperative that all samples, collected by whatever means, should be transported as quickly as possible back to the laboratory while holding their temperature as close to that of the collection site as is possible. The latter objective may be achieved by the use of 'thermos' flasks, or by transporting the samples packed in polystyrene foam beds or blocks, or by submerging them in larger volumes of water collected on site. After reaching the laboratory the collections should be placed in some form of dish that has a large surface area to volume ratio, this will not only help keep the samples aerobic but will also allow easy examination under a dissecting microscope. However, samples collected for anaerobic ciliates should not be treated in this manner, they should be kept in closed vessels which have a low surface area to volume ratio and preferably should contain large amounts of organic debris. The presence of debris and associated bacteria will serve to keep the level of dissolved oxygen to a minimum. In all cases the temperature should be maintained as near as is possible to that of the collection site until the ciliates are either identified or cultured.

Raw samples obtained from the field will contain a mixture of organisms which are not under control and as time goes by the population will change

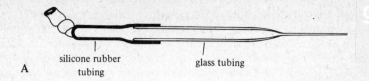

A   silicone rubber tubing   glass tubing

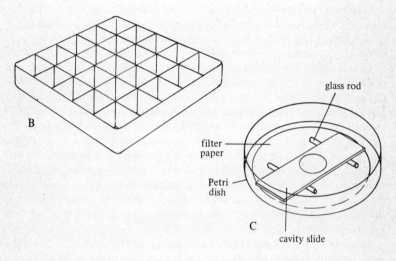

B

glass rod

filter paper

Petri dish

C

cavity slide

Fig. 13. Apparatus used for the isolation and culture of ciliates. A, Micropipette (half size). B, Divided Petri dish. C, Moist chamber.

radically both in terms of numbers of individuals and species present. Samples such as these may be enriched by the addition of liquid culture media of various kinds (see below) or by adding cracked grains of wheat, rice or other cereal seeds. Additions such as these generally raise the numbers of certain ciliates present (by raising the bacterial population present) but often lower the number of species. However, rough cultures containing several ciliate species and unknown bacterial and algal floras can be maintained for several months and sometimes years. A greater degree of control over a culture may be obtained by picking out a particular ciliate species and culturing it in the isolation of organisms other than its prey. This procedure is generally carried out under a low-power (10–75 times magnif.) stereoscopic dissecting microscope using a micropipette. The latter are generally constructed in pairs from a piece of glass tubing (8–9 cm long, 4 mm diameter) which is heated in the centre until red hot, it may then be pulled out such that finely narrowed tips are formed. After cooling, the two pipettes are separated and the wide ends are flame polished to eliminate sharp edges. Rubber teats normally available from laboratory suppliers and chemists are generally too large but smaller ones may easily be constructed from 4 mm silicone rubber tubing with a knot tied at one end, the correct working volume of the teat will be achieved by using 3–4 cm of tubing (see Fig. 13A). It takes only a few hours practice to become adept at handling ciliate cells using micropipettes and with a little more practice one can easily pick out single cells down to a diameter of 30 μm or less. Cultures of single species may be initiated by inoculating several individuals of what appear to be the same species into the culture medium but really the only way to be certain that the culture contains a single species is to inoculate it with a single cell. A population arising from a single cell by asexual binary fission is known as a **clone**. It is advisable, when starting clonal cultures, to set up several simultaneously. This will prevent a certain amount of frustration and time since cells are frequently damaged during transfers and do not grow. Plastic dishes that are divided into many small compartments (Fig. 13B) are ideal for this purpose but cavity slides kept in moist chambers will suffice (Fig. 13C). After a day or so, the cultures should be examined and successful ones may then be transferred to larger vessels containing fresh culture media. Glass and plastic Petri dishes (9 and 5 cm in diameter) are perhaps the best containers for the maintenance of cultures but really almost any glass or plastic vessel could be used. Cultures will eventually die unless transferred to fresh culture media, the frequency of transfer depends to a large extent upon the ciliate and its medium but normally this should be carried out at 2–3 week intervals.

The choice of a culture medium will depend largely upon what the ciliate feeds. Many feed upon bacteria and this is why many of the media commonly used are designed to encourage the growth of bacterial populations. These media are frequently weak infusions of natural materials which promote the growth of sufficient bacterial numbers for food without reducing the concentration of dissolved oxygen too far. The three most commonly used culture media are outlined in Table 1 but the reader is recommended to consult the culture media and methods published in the following works, Mackinnon and Hawes (1961), Kirby (1950) and Committee on Cultures, Society of Protozoologists (1958).

Cultures should be examined under a low-power stereoscopic dissecting microscope taking care to focus at all levels in the culture watching for movement of any kind. Ciliates for closer examination and identification should be picked out with a micropipette, mounted on a glass slide and viewed with a good quality high-power (64–1500 times magnif.) microscope. It is a great advantage if the microscope is equipped with a mechanical stage and phase contrast illumination. Cover slips may be placed directly onto the drop of water containing the ciliates but if they are required to be active for several hours the cover slip should be supported by a ring of vaseline.

The rapid movement of ciliates often makes it difficult to observe them properly; tracking them using a mechanical stage helps considerably but sometimes one has to resort to using some method to slow or immobilise the

Table 1. *Three commonly used culture media for ciliated protozoa*

	Weight used per litre distilled water	Method of preparation
Soil	5 g garden top soil	Boil 2 hours, settle overnight, decant and filter supernatant
Hay	1–2 g Timothy hay or similar	Boil, filter
Lettuce	1–2 g dried lettuce leaves	Boil, filter

organisms. Physical techniques are less drastic than chemical ones which eventually kill the animal. Larger ciliates may be trapped in small areas on the slide by mounting the ciliates on top of a small amount of teased-out cotton wool. The fibres of the cotton wool restrict the movement of the cells to a small area without affecting them adversely. The use of methyl cellulose to increase the viscosity of the mounting fluid has often been used to slow the swimming speed of a ciliate. The solution is prepared by dissolving 10 g methyl cellulose in 45 ml boiling water. It should be allowed to stand for 20–30 minutes, diluted with an equal amount of cold water and mixed to a smooth paste. A ring of methyl cellulose is made on the slide and a drop of the water containing ciliates is placed in its centre. A cover slip is added and the ciliates slow down progressively as the methyl cellulose diffuses inward. Chemical means of immobilisation are often suggested but it should be stressed that different ciliates tend to react differently to the same chemical and choice is purely empirical. The following solutions have been suggested; 1% copper sulphate, 1% nickel sulphate, saturated sodium amytal (needs 20–30 minutes to react) and 10% methyl alcohol. In all cases the ciliates should be mounted in the normal way and the immobilisation fluid is added to one edge of the preparation so that it seeps under the cover slip and diffuses across. Using this method a concentration gradient will be set up across the slide and usually the ciliates in one area will be slowed or immobilised without immediate death or bursting.

Several stains may be used on living or freshly mixed animals to display organelles that may be difficult to see but which are important for identification purposes. Nuclei may be stained by adding a drop of a 1% solution of methyl green in 1% acetic acid to a drop of the culture. This solution both fixes the animal and stains the nuclei green. Cilia and cirri may be stained either with toluidine blue or with Noland's stain. In the former method the ciliates are first fixed by inverting a drop of the culture over a 1% solution of osmium tetroxide for 10–20 seconds. This procedure should be carried out in a fume cupboard since osmic acid fumes are highly toxic. A drop of 0.01% aqueous solution of toluidine blue is then added to the fixed cells and covered with a cover slip. Cilia, cirri and membranelles stain an intense blue. Noland's stain contains 80 ml of a saturated solution of phenol in distilled water, 20 ml formalin (40% aqueous solution of formaldehyde), 4 ml glycerol and 20 mg gentian violet. The latter ingredient is the dye which should be moistened with a little water before adding the other ingredients and it is important that the final solution does not contain phenol in suspension. A drop of the combined fixative and stain is added to a drop of the culture.

The above methods are often sufficient for most identification purposes but at some stage it may be necessary to make permanent preparations of ciliates. Nuclei may be selectively stained using the Feulgen reaction (for method see Mackinnon and Hawes, 1961) and it will be found necessary to use silver impregnation methods to differentiate between certain genera. In general the key has been arranged such that the necessity of silver staining is kept to an absolute minimum but in some cases there is no other method of positive identification.

There are three common methods of staining the infraciliature of ciliates, the 'dry' method of Klein, the 'wet' method of Chatton and Lwoff and the protargol method. All three methods have advantages and disadvantages and it should be emphasised that none of these methods, particularly the 'wet' method, is always successful on all ciliate species. The ultimate choice is based on trial and error. When a silver method needs to be used in the key an indication of which methods have been successfully used in the past will be given.

The 'dry' silver method of Klein (1958) is in our experience the quickest and easiest method to use, but frequently the cell distorts during fixation and incidently this method cannot be used for staining marine ciliates. The cells are picked out with a micropipette and placed in a series of tiny drops of water on a polished, grease-free slide (previously cleaned in ether). Excess water must be removed by means of a micropipette or with small screws of tissue paper and the slide allowed to dry in the air; the cells should die and dry simultaneously for good fixation. The dry slide is then submerged in 2% silver nitrate solution for 15 minutes at room temperature, followed by a quick wash in distilled water. The slide is then exposed for at least 15 minutes to a UV lamp or strong sunlight. The correct exposure being determined by frequent inspection under a microscope. When the infraciliature turns black, the preparation is washed in tap water and air dried in a vertical position. The preparation may then be cleared in xylol and mounted in Canada Balsam or equivalent mounting medium.

The 'wet' silver-impregnation method was introduced by Chatton and Lwoff (1930) and modified slightly by Corliss (1953). The following scheme is based on our own experience.

1. Centrifuge culture and remove supernatant with a pipette.
2. Add Champy's fluid and leave for 3–5 minutes only. [Champy's fluid – 7 ml chromic acid (1% aq), 7 ml potassium bichromate (3% aq) and 4 ml osmium tetroxide (2% aq) – should be mixed freshly *immediately* before use by the addition of the final component, osmic acid, to a mixture of the other two. Failure to do this is one of the most common causes of non-success with this method.]
3. Add da Fano's fixative, centrifuge, remove supernatant, add more da Fano's fixative. This should be repeated if the ciliates are too yellow in colour. Specimens may be stored in this fixative for several weeks. [Da Fano's fixative consists of 1 g cobalt nitrate, 1 g sodium chloride, 10 ml formalin, dissolved in 90 ml distilled water.]
4. Ciliates in the final da Fano wash should be poured into a glass dish and left in a refrigerator for at least 2 hours. During this time, trays of ice should be prepared and bottles of distilled water placed in the refrigerator to cool along with a moist chamber (see Fig. 13C). Finally, saline gelatin, glass rods and clean microscope slides should be warmed (35–45 °C) on a hot plate or water bath. [Saline gelatin is prepared from 10 g powdered gelatin, 0.05 g sodium chloride dissolved in 100 ml distilled water.]
5. Put a drop of the culture onto a warm slide and add an equal volume of warm saline gelatin. Spread with a warm glass rod and *immediately* place in the cold moist chamber which is kept on a bed of ice. After several minutes the gelatin will have set and it may then be washed with a jet of cold distilled water.
6. Flood slide with cold 3% silver nitrate and leave in cold and dark for 20–30 minutes.
7. Wash in jet of cold distilled water and then place in Petri dish, cover with cold distilled water and keep on a bed of ice. Expose to UV lamp (CARE should be taken to protect the eyes and skin) or preferably to sunlight for 20–30 minutes. Determine correct exposure time by inspection under a microscope.
8. Remove to 70% alcohol, dehydrate through alcohols, clear in xylol and mount in Canada Balsam.

Care should be taken that the gelatin/ciliate preparation is not too thick as this can result in the gelatin peeling off in the alcohols or it may prove difficult to dehydrate. Occasionally problems with dehydration may be encountered (the gelatin turns milky in xylol), this may be overcome by further dehydration in fresh absolute alcohol. If this is unsuccessful clear in clove oil.

The protargol (protein silver) method was first developed for staining vertebrate nervous tissue but more recently has been successfully used for staining the infraciliature of ciliates. Certain workers have had difficulty in applying this method successfully. There are several variations, Tuffrau (1967) for example published his method in detail. The method described below is the one we have found to be reliable for us. The procedure should be carried out on cover slips rather than slides since this reduces the amounts of expensive chemicals required.

1. Centrifuge culture to concentrate ciliates, remove supernatant with pipette.
2. Fix cells for 5–10 minutes in *fresh* Bouin fixative [prepared from 75 ml saturated solution of picric acid, 25 ml formalin and 5 ml glacial acetic acid]. *Do not* keep cells in this fixative for long periods.
3. Dehydrate in 70% and absolute alcohols. It is not necessary to remove all the fixative during this procedure.
4. Spread albumen/glycerin onto coverslips, dry on a hot plate (45–50 °C) until just sticky. Add a drop of the culture and leave on hot plate for 1–2 minutes to evaporate off the alcohol and glycerin. Cover with formol/alcohol and leave 5–30 minutes to set the albumen. [Albumen/glycerin may be freshly prepared by mixing 2 parts of egg white with 1 part glycerin by volume or it may be bought from laboratory suppliers as Mayer's albumen. Formol/alcohol is prepared from absolute alcohol and formalin in equal volumes.]
5. The cells on the coverslip are then dehydrated for 5 minutes in each of the following alcohols, 95%, twice in absolute ethyl alcohol and finally in absolute methyl alcohol.
6. The coverslip is then briefly dipped into 0.5% parlodion dissolved in methyl alcohol, until it has a chalky white appearance.
7. Rehydrate for 5-minute periods in 70%, 50% and 30% alcohol followed by three quick (1–2 minute) washes in distilled water. Longer immersion in water tends to dislodge the albumen.
8. Place into fresh 0.5% potassium permanganate for 5 minutes. The cells will turn brown.
9. Give two quick washes in distilled water and submerge in 3–5% oxalic acid. The cells should be left until they have changed from brown to greyish-white.
10. Wash thoroughly in three changes of distilled water. Stain for 1 or 2 days in protargol at room temperature. This time may be reduced to $\frac{1}{2}$–2 hours by raising the temperature of the protargol (up to 80 °C). Times and temperatures are estimated by trial and error. Protargol is known as strong silver proteinate or silver albuminose. [It is prepared by sprinkling 1 g Roques protargol over 100 ml of distilled water. Leave undisturbed to dissolve, do not stir. When solution is complete, pour into staining trough containing small pieces of copper wire.]

11. After staining briefly wash twice in distilled water.
12. 'Develop' the silver by immersion in a solution of 1 g hydroquinone dissolved in 100 ml of 5% sodium sulphite solution. The length of time in this solution varies with staining time, the shorter the staining period the longer the development period should be. The development procedure should be observed under a low power dissecting microscope. On development the nuclei should be dark brown and the cytoplasm will be a much lighter shade of brown.
13. Tone in 1% gold chloride until the preparation turns purple (usually 2–5 minutes) then wash three times in distilled water.
14. Fix in 5% sodium thiosulphate for 5–10 minutes. Wash twice in distilled water.
15. Dehydrate slowly through the alcohols, clear in xylol and mount.

Recently Fernandez-Galiano (1976) introduced another wet method that several authors have used with success. This method stains certain ciliary organelles rather better than any other technique available but has the disadvantage of producing impermanent slides. However, the results are said to be good enough to produce excellent photomicrographs as a permanent record. The other disadvantage with this method is that one step using pyridine should be carried out in a fume cupboard which is rarely available to the amateur.

Keys to certain groups of ciliated protozoa

The keys published here have been compiled with the user in mind, we have not followed the usual taxonomic approach since many of the features required to be seen in such keys are not always easily visible. Furthermore it is not normally possible to give a short diagnostic clue that will immediately place all relevant genera into the correct subclass, then order and so on, there are too many exceptions and too many situations where the taxonomic location of certain genera is imprecisely known because of the lack of data. Furthermore taxonomy is always in a state of flux and such keys quickly become dated.

Our approach has been to use non-taxonomic criteria first to reduce the numbers of genera to smaller sized groups. For example, the most obvious feature one sees when looking at a ciliate is its habit. The observer immediately knows if the animal concerned is free-swimming, sessile or loricate; such features do not place the organism into any neat taxonomic grouping but do reduce the number of genera needed to be considered. For this reason the first key separates loricate, free-swimming, sessile and suctorian genera. Of these only the suctoria involve a single taxonomic group; the other three groupings are then treated individually to divide their component genera into their appropriate subclass and finally their genus. In most cases division into a particular subclass is performed in several stages since in practice all genera belonging to a subclass cannot usually be placed there by the application of a single short diagnosis.

Generic descriptions are arranged taxonomically according to the classification given by Corliss (1979). The actual taxonomic grouping chosen varies from subclass to subclass (usually order and/or suborder but occasionally family) but the genera are listed alphabetically within these groupings. In certain cases we have found that we could not agree with Corliss (1979) over certain matters, and/or the situation had changed since that publication went to press, in such cases minor variations may be found but we usually point out the discrepancy and our reasons.

1. Adult always with cilia and/or cirri, when tentacle-like processes are present they are not hollow **2**

Adult never with cilia or cirri, hollow suctorial tentacles always present .... SUCTORIA (p. 298)

2. Ciliate either partly or totally encased in some form of lorica or gelatinous housing .... LORICATE GENERA (p. 32)

Ciliate not in a case **3**

3. Ciliate not fixed to a substratum, but swimming freely in water or moving over the surface of a substratum. Never possesses a stalk .... FREE-SWIMMING GENERA (p. 36)

Ciliate attached to a substratum (animal, plant or inanimate object) by some means. Some possess a stalk ...... SESSILE GENERA (p. 58)

Key to subclasses of loricate ciliates

1. Cirri present, cilia also usually present but these may be difficult to observe .... LORICATE SPIROTRICHA (Part II)

   Cirri absent, cilia present **2**

2. Cilia restricted to anterior body quarter **3**

   Cilia distributed all over body **4**

3. Cilia restricted to apical region to form an AZM (p. 8) which winds anticlockwise to the cytostome. Lorica may be pseudo-chitinous or mucilaginous but never extensively constructed with sandgrains and other debris (Fig. 14A) ....
   .... LORICATE PERITRICHA (Part II)

   Cilia may be in anterior body quarter and/or in apical region. Apical cilia form an extensive AZM which winds clockwise to the cytostome. Lorica often extensively constructed of sandgrains and other debris (Fig. 14B) .... LORICATE SPIROTRICHA (Part II)

4. Lorica upright, rarely lying on side. There are never wing-like extensions to anterior region of body **5**

   Lorica lying partly on side, AZM on edges of two wing-like anterior extensions to body (Fig. 14C) ....
   .... LORICATE SPIROTRICHA (Part II)

5. Lorica pseudochitinous or mucilaginous, never constructed with sandgrains **6**

   Lorica extensively constructed of sandgrains and other debris .....
   .... LORICATE SPIROTRICHA (Part II)

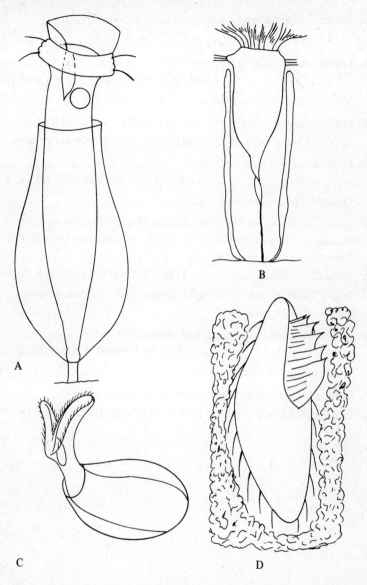

Fig. 14. Some loricate ciliates. A, *Cothurnia*. B, *Tintinnidium*. C, *Diafolliculina*. D, *Bothrostoma*.

6. Lorica open at one end **9**

Lorica open at both ends **7**

7. Lorica a simple tube with parallel sides ....
.... LORICATE HYMENOSTOMATA (Part II)

Sides of lorica never parallel **8**

8. Lorica widest in centre .. LORICATE HYMENOSTOMATA (Part II)

Lorica widest at one end ... LORICATE GYMNOSTOMATA (p. 61)

9. Lorica forming a branched colony ....
.... LORICATE VESTIBULIFERA (p. 176)

Lorica unbranched, solitary **10**

10. Conspicuous large AZM either apically (Fig. 19B) or on most of
the anterior half of the body (Fig. 14D). In both cases the AZM
winds clockwise to the cytostome ....
.... LORICATE SPIROTRICHA (Part II)

When AZM present it is never conspicuous and never winds
clockwise to the cytostome **11**

11. Lorica pseudochitinous (membranous and rigid) ....
.... LORICATE GYMNOSTOMATA (p. 61)

Lorica gelatinous or mucilaginous **12**

12. Contractile vacuole terminal or in posterior region of body **13**

Contractile vacuole apical ... LORICATE VESTIBULIFERA (p. 176)

13. Contractile vacuole always terminal, mouth apical without membranelles .... LORICATE GYMNOSTOMATA (p. 61)

Contractile vacuole usually posterior but rarely terminal, oral aperture anterior but lateral with short membranelles ....
.... LORICATE VESTIBULIFERA (p. 176)

Key to subclasses of free-swimming ciliates

1. Cirri present (NB in one rare small hymenostome genus *Sagittaria*, thickened stiff cilia could be mistaken for cirri, Fig. 20A) **2**

   Cirri absent **6**

2. Cirri in short rows, lines or groups, when they encircle the body they do so spirally not transversally (Fig. 15A) **3**

   Cirri in the form of bristles which form one or more transverse bands to encircle the body but never spirally so (Fig. 15B,C) **5**

3. Cirri restricted to single caudal group **4**

   Cirri distributed over body, when in small groups never just restricted to single caudal group ....
   .... FREE-SWIMMING SPIROTRICHA (Part II)

4. With spines or spikes, laterally compressed body with rigid pellicle. Reduced somatic ciliation. In anaerobic habitats ....
   .... FREE-SWIMMING SPIROTRICHA (Part II)

   Without spines and spikes, body rounded in cross-section. Many somatic cilia. Planktonic, rare (*Sulcigera*) ....
   .... FREE-SWIMMING VESTIBULIFERA (p. 177)

5. Single row of bristle-like cirri encircles body. Apical rounded oral aperture without AZM (Fig. 15B). Cytopharynx often supported by prominent basket of trichites ....
   .... FREE-SWIMMING GYMNOSTOMATA (p. 62)

   Several rows of bristle-like cirri encircle body. Peristomial area displaced to one side with AZM leading to cytostome which is never supported by a basket of trichites (Fig. 15C) ....
   .... FREE-SWIMMING SPIROTRICHA (Part II)

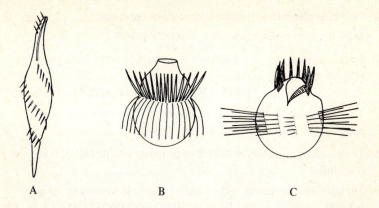

Fig. 15. Some ciliates with cirri. A, *Urostrongylum*, a spirotrich with spiralling cirri. B, *Mesodinium*, a gymnostome with a single transverse band of cirri. C, *Halteria*, a spirotrich with several transverse bands of cirri.

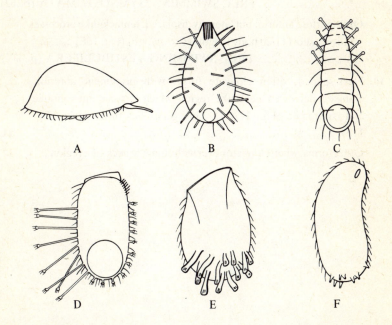

Fig. 16. Tentacle-like processes on some non-suctorian ciliates. A, *Lophophorina*. B, *Actinobolina*. C, *Enchelyomorpha*. D, *Thysanomorpha*, with tentacles extended on left. E, *Legendrea*. F, *Cirrophrya*.

6. Only cilia present, no tentacle-like processes present **10**

Cilia and tentacle-like processes present (Fig. 16A,F) **7**

7. Single, mobile, anterior tentacle-like process present. Rare genus
(*Lophophorina* Fig. 16A) ....
.... FREE-SWIMMING HYPOSTOMATA (p. 225)

Two or more tentacle-like processes present which may be
anteriorly or posteriorly situated (Fig. 16B–F) **8**

8. Tentacle-like processes restricted to posterior end or half of body
(Fig. 16E–F) **9**

Tentacle-like processes either anterior or all over body (Fig.
16B–D) .... FREE-SWIMMING GYMNOSTOMATA (p. 62)

9. Oral aperture apical, tentacle-like process relatively long (1/3–1/8
body length) (Fig. 16E) ....
.... FREE-SWIMMING GYMNOSTOMATA (p. 62)

Oral aperture anterior but laterally displaced, tentacle-like processes
relatively short (1/20 body length) (*Cirrophyra*) (Fig. 16F) ....
.... FREE-SWIMMING VESTIBULIFERA (p. 177)

10. Flask-shaped, often elongate forms with prominent anterior,
elongate narrowed neck-like region which is sometimes mobile
and contractile (Fig. 17A–I). One genus has three necks (Fig.
17F) **11**

Other forms, when elongate never with elongate neck-like region ..   **12**

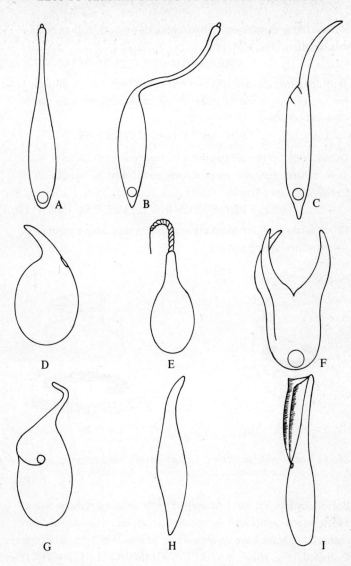

Fig. 17. Outline shapes of some ciliates with anterior neck-like regions. A, *Trachelophyllum*. B, *Lacrymaria*. C, *Dileptus*. D, *Trachelius*. E, *Ileonema*. F, *Teuthophrys*. G, *Paradileptus*. H, *Amphileptus*. I, *Cohnilembus*.

11. Without long membranelles stretching down neck region, without caudal cilium (Fig. 17A–H) ....
.... FREE-SWIMMING GYMNOSTOMATA (p. 62)

With long membranelles (two present but one is more difficult to see) stretching longitudinally down neck region, with caudal cilium (*Cohnilembus*) (Fig. 17I) ....
.... FREE-SWIMMING HYMENOSTOMATA (Part II)

12. Ovoid body, many contractile vacuoles scattered all over body, large central vacuole, elongate macronucleus, no mouth. Rare free-living form (*Archiastomata*) ....
.... FREE-SWIMMING HYMENOSTOMATA (Part II)

Other forms without complete combination of above characters, always with a mouth present **13**

A   B

Fig. 18. Peritrichs with an anterior ciliary wreath winding anticlockwise to the cytostome. A, *Astylozoon*. B, *Trichodina*.

13. Barrel, cylindrical or bell-shaped with anterior ciliary wreath which winds anticlockwise to the cytostome. Somatic ciliature reduced but some have an aboral ring of cilia. Macronucleus often C-shaped. (Fig. 18) .... FREE-SWIMMING PERITRICHA (Part II)

When anterior ciliary wreath present it is simple circlet and never winds anticlockwise to cytostome **14**

14. Anterior end and/or left side with conspicuous AZM which winds clockwise to cytostome (Figs. 19 and 20) **15**

When anterior pole has an AZM it is never large and conspicuous and never winds clockwise to cytostome **17**

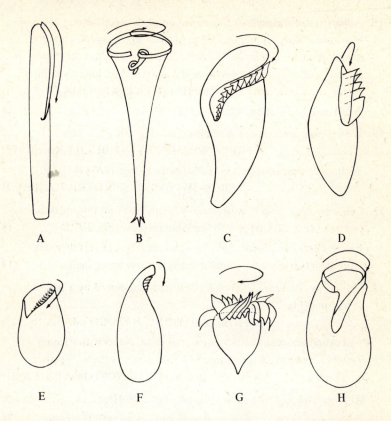

Fig. 19. Ciliates with conspicuous adoral zones of many membranelles which wind clockwise to the cytostome. A, *Spirostomum*. B, *Stentor*. C, *Metopus*. D, *Bothrostoma*. E, *Palmarium*. F, *Tesnospira*. G, *Strobilidium*. H, *Bursaria*.

15. Apparent conspicuous AZM really either a membranous collar (rare genus *Sagittaria*, Fig. 20A) or made up of paired cilia not membranelles (*Neobursaridium*, Fig. 20B). In the latter case there are two contractile vacuoles each served by several long radiating canals .... FREE-SWIMMING HYMENOSTOMATA (Part II)

AZM made up of membranelles (Fig. 19A–H) **16**

16. With conspicuous cleft in ventral surface (Fig. 20C) ....
.... FREE-SWIMMING VESTIBULIFERA (p. 177)

Without conspicuous cleft in ventral surface (Fig. 19A–G) ....
.... FREE-SWIMMING SPIROTRICHA (Part II)

17. Cell with single (rarely two), usually blunt, posterior protoplasmic processes (Fig. 21A,B) or with single lateral spine (Fig. 21C,D) ...   **18**

Processes may be absent, but when present (Fig. 21E–H) there are more than two and are accompanied by sharp anterior spines **19**

18. Cell always flattened, cytopharynx usually supported by a basket of trichites (Fig. 21A–D) ....
.... FREE-SWIMMING HYPOSTOMATA (p. 225)

Cell cylindrical with shallow equatorial waist. Never with trichites. Tail-like process is a clump of cilia which adhere together (*Urocentrum*) ... FREE-SWIMMING HYMENOSTOMATA (Part II)

19. Anterior and posterior spines present (Fig. 21E–H) **20**

Posterior spines never present. Anterior spines very rarely present   **21**

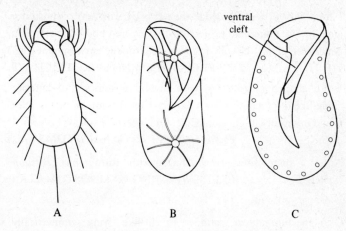

Fig. 20. Some ciliates with oral structures easily mistaken for adoral zones of membranelles. A, *Sagittaria*. B, *Neobursaridium*. C, *Bursaria*.

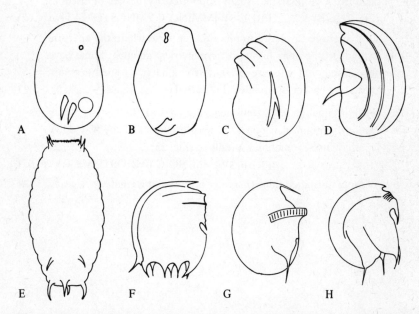

Fig. 21. Some ciliates with spine-like projections. A, *Parachilodonella*. B, *Dysteria*. C, *Microthorax*. D, *Drepanomonas*. E, *Coleps*. F, *Saprodinium*. G, *Discomorphella*. H, *Atopodinium*.

20. Laterally compressed, wedge-shaped body with several posterior and anterior spine-like projections. Armoured body often ribbed, never with plates. Usually found in a sapropelic environment (Fig. 21F–H) .... FREE-SWIMMING SPIROTRICHA (Part II)

Barrel-shaped (Fig. 21E), with several posterior and anterior spine-like projections. Body covered in armour of plates. Not found in sapropelic environments. Oral aperture apical (*Coleps*) .... FREE-SWIMMING GYMNOSTOMATA (p. 62)

21. Anterior spines present encircling mouth. Rare genus (*Pseudo-enchelys*) .... FREE-SWIMMING GYMNOSTOMATA (p. 62)

No spines present **22**

22. Oral aperture apical, rounded or slit-like, lying symmetrically about the apex which may be obliquely truncated. Often with somatic cilia around oral aperture but never with specialised oral ciliature. Cytopharynx usually supported by basket of trichites (Fig. 22A–E) .... FREE-SWIMMING GYMNOSTOMATA (p. 62)

Oral aperture displaced to one side, in those where the aperture extends to the apex it is never symmetrical about it. Some have specialised oral ciliature present. Cytopharynx sometimes supported by basket of trichites (Fig. 22F–H)............................ **23**

23. With prominent unciliated ridge armed with trichocysts which extend from apical region to at least a quarter of the way down length of body, sometimes spirally so (Fig. 23) .... FREE-SWIMMING GYMNOSTOMATA (p. 62)

Without unciliated ridge (if a ridge is present it is ciliated) **24**

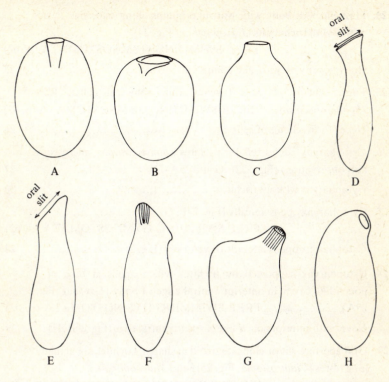

Fig. 22. Apical oral structures in ciliates. Those represented in A–E are considered to be symmetrical about the body apex. Those in F–H are considered to be unsymmetrical about the apex. A, *Holophrya*. B, *Bursellopsis*. C, *Monodinium*. D, *Spathidium*. E, *Acineria*. F, *Cranotheridium*. G, *Spathidioides* Brodsky. H, *Woodruffia*.

24. Flask-shaped, body with wart-like bumps along one edge, warts armed with trichocysts (*Loxophyllum*, Fig. 24) ....
.... FREE SWIMMING GYMNOSTOMATA (p. 62)

Body never with wart-like bumps **25**

25. Forms chains of cells (usually four spherical cells). Rare (*Sphaerobactrum*).....FREE-SWIMMING GYMNOSTOMATA (p. 62)

Never forms chains of cells **26**

26. Cytopharynx supported by trichites (phase-contrast microscopy aids observation) (Figs. 22F,G and 25A–H) **27**

Cytopharynx without trichites **30**

27. Cytopharynx opens apically (Fig. 22F, G) ....
.... FREE-SWIMMING GYMNOSTOMATA (p. 62)

Cytopharynx opens laterally or ventrally (Fig. 25A–H) **28**

28. Cytopharynx supported by trichites which opens at base of a pouch-like cavity in anterior ventral edge of body (*Loxodes*, Fig. 25A) .... FREE-SWIMMING GYMNOSTOMATA (p. 62)

Never with anterior pouch-like cavity in ventral edge (Fig. 25B–H)   **29**

29. Oral aperture slit or elongate ovoid in shape. Trichites are actually long fibres (*Clathrostoma*, Fig. 25H and *Malacophrys*) ....
.... FREE-SWIMMING HYMENOSTOMATA (Part II)

Other forms, cytopharynx supported by proper trichites not fibres (Fig. 25B–G) .... FREE-SWIMMING HYPOSTOMATA (p. 225)

30. Closely associated with exuvial fluid of moults of crustacea **31**

Not closely associated with crustacean moults **32**

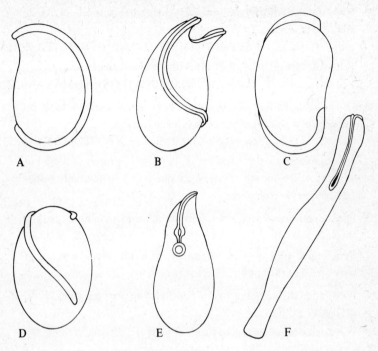

Fig. 23. Some ciliates with unciliated ridges. A, *Bryophyllum*. B, *Diceratula*. C, *Penardiella*. D, *Perispira*. E, *Branchioecetes*. F, *Myriokaryon*.

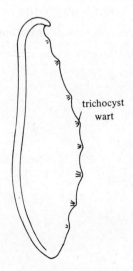

Fig. 24. *Loxophyllum*, a ciliate with trichocyst warts.

31. Anterior end with forward projecting cilium, no rosette organelle (Fig. 26) (*Larvulina*) ....
.... FREE-SWIMMING HYMENOSTOMATA (Part II)

With rosette organelle (Fig. 26), without apical cilium ....
.... FREE-SWIMMING HYPOSTOMATA (p. 225)

32. Dorso-ventrally flattened. Oral aperture, a transverse wide slit on the ventral surface (*Gastronauta*) ....
.... FREE-SWIMMING HYPOSTOMATA (p. 225)

If dorso-ventrally flattened the oral aperture is rarely slit-like but when it is, the mouth is either obliquely or longitudinally orientated but never transverse **33**

33. Peristomial area extends to apex of cell and sometimes over it (Figs. 27 and 28) **34**

Peristomial area may be lateral or posterior, when anterior it never reaches the apex **39**

34. Peristomial area extends over or encircles apex of cell (Figs. 20A, 22H and 27A–F) **35**

Peristomial area just reaches apex of body (Fig. 28) **37**

35. Peristomial area restricted to apical region, does not extend down side of cell (Figs. 22H and 27B,C,E)......................................
.............................FREE-SWIMMING VESTIBULIFERA (p. 177)

Peristomial area extends down side of cell but touches apex (Figs. 20A and 27A,D,F) **36**

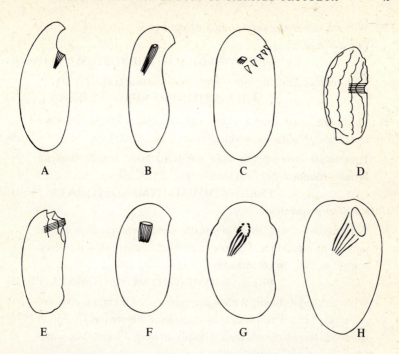

Fig. 25. Position of trichites and trichite-like structures in certain ciliates. A, *Loxodes*. B, *Synhymenia*. C, *Nassula*. D, *Drepanomonas*. E, *Stammeridium*. F, *Chilodonella*. G, *Chlamydodon*. H, *Clathrostoma* with fibrils supporting the cytopharynx.

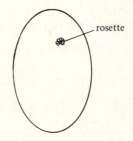

Fig. 26. The rosette organelle in apostome hypostomes.

36. With membranelle surrounding or emerging from oral aperture
    (Figs. 20A, 27F) ....
    .... FREE-SWIMMING HYMENOSTOMATA (Part II)

    Without membranelle surrounding oral aperture (Fig. 27A,D) ...
    .... FREE-SWIMMING VESTIBULIFERA (p. 177)

37. Peristomial area approximately one-third body length. Anterior
    end gradually narrows to a blunt point (Fig. 28A,D) **38**

    Peristomial area greater than one-third body length, anterior
    broadly rounded, does not narrow (Fig. 28B,C,E) ....
    .... FREE-SWIMMING HYMENOSTOMATA (Part II)

Make a silver preparation.

38. With caudal cilium, three unequally sized membranelles and an
    undulating membrane. Cytoproct slit-like situated below mouth on
    ventral surface. Never loricate ....
    .... FREE-SWIMMING HYMENOSTOMATA (Part II)

    Without caudal cilium. With four equally sized membranelles and
    two rows of cilia which imitate an undulating membrane. Cytoproct
    rounded, terminally situated. Usually lives in a lorica but some-
    times leaves it (*Cyrtolophosis*) ....
    .... FREE-SWIMMING VESTIBULIFERA (p. 177)

39. With spiral groove restricted to posterior two-thirds of body,
    groove leads to oral aperture ....
    .... FREE-SWIMMING VESTIBULIFERA (p. 177)

    If groove present it begins at apex of cell **40**

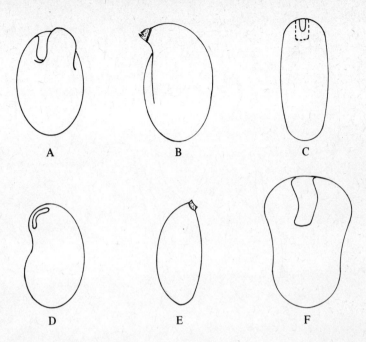

Fig. 27. Some ciliates whose peristomial area extends over or encircles the apex of the cell. A, *Opisthostomatella*. B, *Rigchostoma*. C, *Orcavia*. D, *Bursostoma*. E, *Platyophrya*. F, *Marituja*.

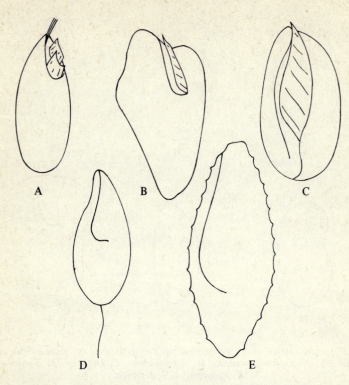

Fig. 28. Some ciliates whose peristomial area just reaches the apex of the cell. A, *Cyrtolophosis*. B, *Parastokesia*. C, *Lembadion*. D, *Paranophrys*. E, *Ctedoctema*.

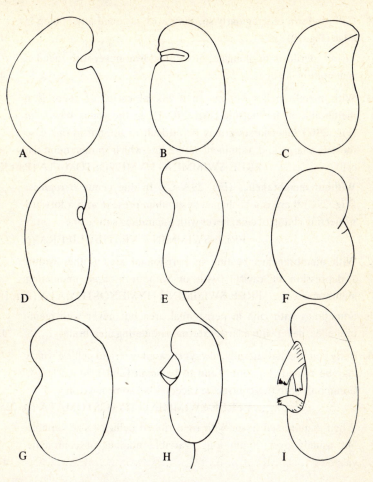

Fig. 29. Some reniform-shaped ciliates. A, *Colpoda*. B, *Plagiopyla*. C, *Bryophrya*.
D, *Colpidium*. E, *Dexiotrichides*. F, *Chasmatostoma*. G, *Wenrichia*. H, *Uronemopsis*.
I, *Histiobalantium*.

40. Cell reniform or irregularly so, some with rounded indentation in edge (Fig. 29) **41**

Ovoid, pyriform or elongate, never reniform, never with rounded indentation in edge **42**

41. With membranelles beating in a buccal cacity or there is a membrane on the outside (Fig. 29D–I). In the genus *Wenrichia* (Fig. 29G) this character may be difficult to see but in that case there is a large and prominent suture line which spirals round the cell .... FREE-SWIMMING HYMENOSTOMATA (Part II)

Without membranelles (Fig. 29A–C). In one genus *Bryophrya* (Fig. 29C) there is a U-shaped vestibulum present with a forward projecting clump of cilia, never with a spiralling suture line ....
.... FREE-SWIMMING VESTIBULIFERA (p. 177)

42. With membranelles beating in peristomial area (either within a depression or cavity), with or without a large undulating membrane .... FREE-SWIMMING HYMENOSTOMATA (Part II)

Sometimes with cilia in peristomial area but never with membranelles, never with a large external undulating membrane **43**

43. Body compressed laterally, always encased in a rigid pellicle which may be ribbed (Fig. 30). Tend to be small (under 30 μm long). Commonly with two contractile vacuoles but some have one ......
.... FREE-SWIMMING HYPOSTOMATA (p. 225)

When compressed never with rigid ribbed pellicle. Size variable but usually over 30 μm long, variable number of contractile vacuoles **44**

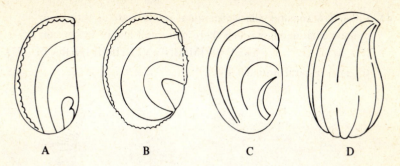

Fig. 30. Some microthoracid hypostome ciliates. A, *Microthorax*. B, *Hemicyclium*. C, *Kreyella*. D, *Trochiliopsis*.

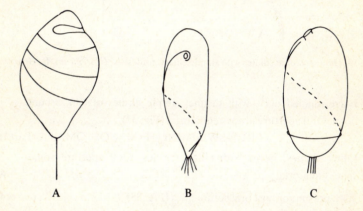

Fig. 31. Trichostomatid vestibuliferans with spiralling cilia. A, *Trimyema*. B, *Spirozona*. C, *Trichospira*.

44. Either few somatic cilia which spiral round body or with many somatic cilia amongst which there is a spiral line of longer cilia (Fig. 31) .... FREE-SWIMMING VESTIBULIFERA (p. 177)

Cilia never spiral around body **45**

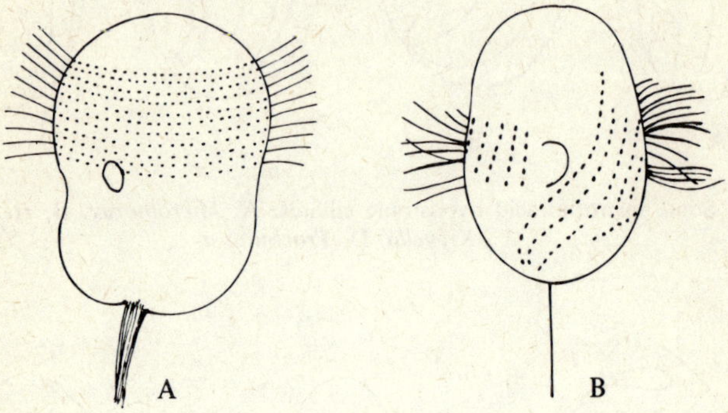

Fig. 32. Hymenostome ciliates with simple ciliary girdles. A, *Urocentrum*. B, *Urozona*.

45. Barrel-shaped body with distinct simple girdles of cilia around body. Oral aperture always equatorial (Fig. 32). ....
.... FREE-SWIMMING HYMENOSTOMATA (Part II)

Other shapes, never with ciliary girdles. Oral aperture sometimes equatorial **46**

46. With pre and/or post oral suture lines (Fig. 33) **47**

Without suture lines ....
.... FREE-SWIMMING GYMNOSTOMATA (p. 62)

47. With oral groove lying along the side or ventral surface which leads to an oral aperture in central region of body (Fig. 33B) **48**

Without distinct oral groove ....
.... FREE-SWIMMING HYMENOSTOMATA (Part II)

48. Ovoid, dorso-ventrally flattened, length less than twice width of body. Single contractile vacuole (*Kalometopia*) (Fig. 33E) ....
.... FREE-SWIMMING VESTIBULIFERA (p. 177)

Cigar, foot-shaped or clavate, not dorso-ventrally flattened, length more than twice its width. Often with two contractile vacuoles (Fig. 33B) ....
.... FREE-SWIMMING HYMENOSTOMATA (Part II)

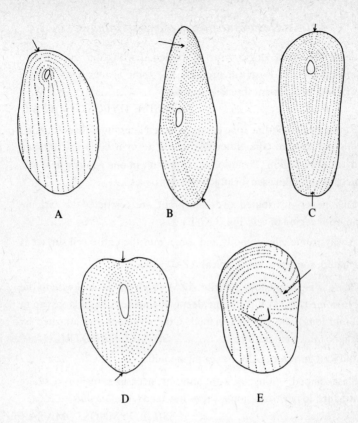

Fig. 33. Some ciliates with suture lines. In each case arrows indicate the positions of preoral suture lines and postoral ones where they exist. A, *Ophryoglena*. B, *Paramecium*. C, *Frontonia*. D, *Stokesia*. E, *Kalometopia*.

Key to subclasses of sessile ciliates

1. Oval in outline, dorso-ventrally flattened with cilia restricted to ventral surface. Posterior protoplasmic spine always present from which an attachment thread is secreted ....
.... SESSILE HYPOSTOMATA (p. 232)

   Rarely oval in outline, one genus flattened laterally, others rounded in cross-section, cilia either distributed all over body or restricted to anterior region. Posterior spine present in one genus where it is a cirrus and is not used for attachment purposes   2

2. Cilia never distributed over body but are restricted to extreme anterior region of cell (Fig. 34C–E)   5

   Always some cilia on body and sometimes the entire cell surface is ciliated, some possess an apical AZM   3

3. There is a conspicuous anterior AZM which winds clockwise to the cytosome (Fig. 34A,B). Some deeply pigmented, often large (up to 2 mm long) and trumpet-shaped. Others smaller and attached by fragile mucilaginous thread .... SESSILE SPIROTRICHA (Part II)

   Without anterior AZM. Never pigmented   4

4. Flask-shaped, elongate with anterior neck-like region. Usually attached to peritrichs upon which they feed (*Amphileptus*) ....
.... SESSILE GYMNOSTOMATA (p. 75)

   Pyriform, without neck region. Attached by posterior stalk-like structure. Never epizooic on peritrichs (*Grandoria*) ....
.... SESSILE VESTIBULIFERA (p. 181)

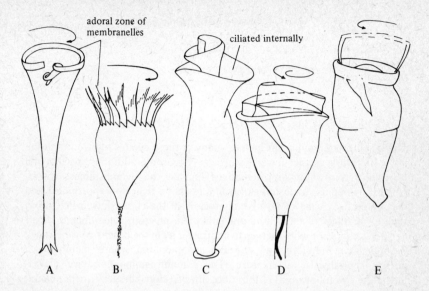

adoral zone of
membranelles

ciliated internally

A   B·   C   D   E

Fig. 34. Ciliary organelles in some sessile ciliates. A, *Stentor*, a spirotrich.
B, *Strobilidium*, a spirotrich. C, *Spirochona*, a chonotrich. D, *Vorticella*, a peritrich.
E, *Apiosoma*, a peritrich.

5. Always attached to mouthparts of crustacea by discs never by
stalks, never colonial. Anterior part of body opens out into an
expanded preoral funnel, a distinct spiral structure (Fig. 34C) which
is ciliated internally. Bodies never contractile ....
.... SESSILE HYPOSTOMATA (p. 232)
Attached to animals (including crustacea), plants and inanimate
objects, sometimes by means of a stalk. Sometimes colonial.
Anterior end with conspicuous circle of cilia which wind anticlockwise to the cytostome (Fig. 34D,E) ... SESSILE PERITRICHA (Part II)

Systematic part

Class KINETOFRAGMINOPHORA

The kinetofragminophorans include many of those ciliates referred to in the older literature as the 'lower holotrichs' plus some additions, so that the free-living freshwater representatives include the gymnostomes, trichostomes, colpodids, hypostomes (including chonotrichs and apostomes) and the suctorians. A list of positive characters of the class is given below but their general lack of specialised oral ciliature is an easily remembered, albeit negative, character which serves to contrast it from the other two subclasses, the Oligohymenophora and the Polyhymenophora whose members all possess specialised oral ciliature. The Kinetofragminophora are characterised by the following: (1) there are always short kineties (or fragments) bearing circumoral, atrial or vestibular ciliature in the oral area of the body; (2) an oral atrium or vestibulum is present in some groups and the somatic infraciliature regularly includes kinetodesmata (subpellicular fibres); (3) stomatogenesis is usually telokinetal but there are exceptions in several groups; (4) the oral aperture is often apically located or near the apex and the cytopharynx is typically supported by a basket of trichites; (5) reproduction is by various methods including budding; (6) there are no compound ciliary organelles present.

Keys and descriptions of genera of Gymnostomata

Subclass GYMNOSTOMATA

Most species are without oral ciliature and the oral aperture is usually on or near the surface of the body and located apically, subapically or laterally. The body shape is often simple and covered with uniform somatic ciliation. Toxicysts are common and the cytopharyngeal apparatus is of the rhabdos type.

Key to genera of loricate Gymnostomata

1. Cells ciliated in terminal region 2
   Cells not ciliated in posterior region which has a terminal blister-like vacuole .... *Metacystis* (p. 82)
2. With caudal cilia .... *Vasicola* (p. 86)
   Without caudal cilia 3
3. With unciliated collar-like apical ridge around oral aperture ....
   .... *Enchelydium* (p. 128)
   Without collar around apical oral aperture .... *Pelatractus* (p. 84)

Key to genera of free-swimming Gymnostomata

1. Usually elongate, flask-shaped body, always with pronounced narrow anterior neck region (there are three necks in one genus but there are never two) (Figs. 17, 35) **2**

   Without narrowed anterior neck region (some genera have a very short snout or apical bump) **17**

2. With single anterior neck **3**

   With three anterior necks (Fig. 17F) .... *Teuthophrys* (p. 148)

3. Oral aperture at base of neck region **4**

   Oral aperture is a slit along neck or at apical end of body **6**

4. Body spherical, not elongate, neck region short (Fig. 17D) ....
   .... *Trachelius* (p. 148)

   Body elongate, never spherical, neck region long (Fig. 17C,G) ..   **5**

5. Neck straight or curved but never spiral. Ventral unciliated ridge of neck joins oral aperture at right angles (Figs. 17C and 36B)......
   .... *Dileptus* (p. 144)

   Neck spirals, ventral unciliated ridge of neck joins oral aperture at a tangent (Figs. 17G and 36C) .... *Paradileptus* (p. 146)

6. Oral aperture apical at end of neck **7**

   Oral aperture a slit along edge of neck **14**

7. With long twisted apical flagellum-like organelle (Fig. 17E) ....
   .... *Ileonema* (p. 110)

   Without flagellum **8**

8. Posterior end with two projections (Fig. 35A) ...... *Urochaenia* (p. 120)

   Without posterior projections **9**

9. Short snout-like bump on end of neck region (Fig. 37) **10**

   Without snout-like bump on end of neck region **11**

10. Neck highly extensible, large apical snout on end of neck (Fig. 37A) .... *Lacrymaria* (p. 110)

    Neck not extensible, small apical snout on end of neck (Fig. 37B)
    .... *Trachelophyllum* (p. 120)

11. Apical region terminates obliquely **12**

    Apical region terminates transversely **13**

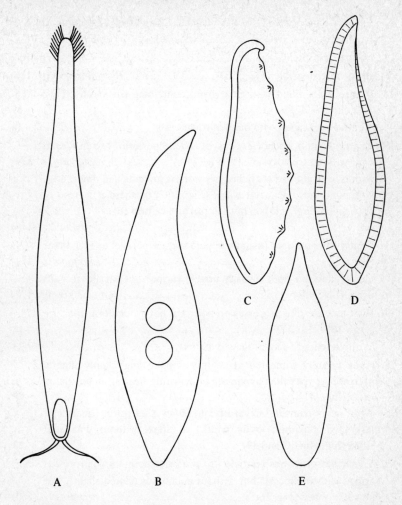

Fig. 35. Some flask-shaped ciliates with anterior neck-line regions. A, *Urochaenia*. B, *Litonotus*. C, *Loxophyllum* with warts. D, *Loxophyllum* with two bands of trichocysts. E, *Rhopalophrya*.

12. Kineties paired apically .... *Protospathidium* (p. 134)
    Kineties never paired .... *Spathidium* (p. 138)
13. Body extensible .... *Chaenea* (p. 104)
    Body not extensible (Fig. 35E) .... *Rhopalophrya* (p. 116)
14. Ventral edge lined with clear band containing trichocysts (Figs.
    23A and 35C,D) **15**
    Ventral edge never with band of trichocysts **16**
15. Ventral band of trichocysts curves around posterior region, dorsal
    edge without trichocyst warts (Fig. 23A) .... *Bryophyllum* (p. 124)
    Either ventral band of trichocysts stops terminally and dorsal edge
    commonly has trichocyst warts (Fig. 35C) or there is a band of
    trichocysts (Fig. 35D) on both dorsal and ventral edges ....
    .... *Loxophyllum* (p. 174)
16. Ciliary rows on right (usually upper) surface parallel to each other
    (Fig. 38A) .... *Litonotus* (p. 174)
    Ciliary rows on right (usually upper) surface not parallel to each
    other (Fig. 38B) .... *Amphileptus* (p. 172)
17. With tentacle-like processes on body but not around oral aperture
    (Figs. 16B–E and 39) **18**
    Without tentacle-like processes on body **23**
18. Apex broadly truncated at slightly oblique angle, only slightly
    narrower at apex than maximum body width, bearing an unciliated
    ridge (Fig. 16D,E) **19**
    Apex never truncate, never oblique, always narrower (quarter or
    less than maximum body width) at apex, without unciliated
    ridge (Figs. 16B,C and 39) **20**
19. Tentacular processes restricted to posterior body third, processes
    armed with trichocysts but without cilia, ends dilated slightly but
    never knobbed (Fig. 16E) .... *Legendrea* (p. 130)
    Tentacular processes restricted to posterior two-thirds of body,
    processes highly extensible armed with trichocysts and ciliated on
    ends which are distinctly knobbed (Fig. 16D) .... *Thysanomorpha* (p. 140)
20. Tentacles distributed all over body **21**
    Tentacles restricted to anterior half of body (Fig. 16C) ....
    .... *Enchelyomorpha* (p. 164)

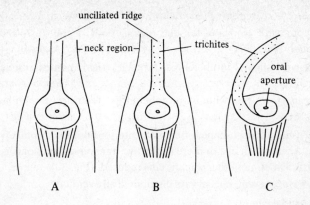

Fig. 36. Oral structures of three gymnostome ciliates. A, *Trachelius*. B, *Dileptus*. C, *Paradileptus*.

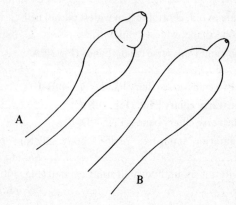

Fig. 37. Position of oral aperture in two ciliates. A, *Lacrymaria*. B, *Trachelophyllum*.

21. Posterior body region rounded, body without furrows **22**

Posterior with elongate, tail-like region, body deeply furrowed obliquely (Fig. 39A) .... *Dactylochlamys* (p. 164)

22. Apex with slight protuberances, tentacles rigid, non-extensible (Fig. 39A) .... *Belonophrya* (p. 162)

Apex narrows but without protuberances, tentacles extensible (Fig. 16B) .... *Actinobolina* (p. 162)

23. Body usually barrel shape, often with anterior nose-like protuberance. Always with one or more distinctive transverse bands of cilia or cirri. Sometimes other somatic cilia reduced. (Fig. 40) **24**

Cirri never present, cilia always distributed all over body **32**

24. With caudal cilium **25**

Without caudal cilium **26**

25. Thickened cilia at apical pole, macronucleus ovoid .... *Peridion* (p. 100)

Circlet of cirri below apical pole, macronucleus moniliform ....

.... *Choanostoma* (p. 154)

26. Approximately ovoid, central region widest part of cell **27**

Short wide cone shape, widest at apex .... *Liliimorpha* (p. 156)

27. With seven to eight transverse ciliary bands (Fig. 40A) ....

.... *Dinophrya* (p. 156)

With one or two transverse ciliary bands (Fig. 40B–F) **28**

28. With one transverse ciliary band (Fig. 40B,C) **29**

With two transverse ciliary bands (Fig. 40D–F) **30**

29. With long anterior snout-like region, body ciliature reduced (Fig. 40B) .... *Monodinium* (p. 160)

With short anterior snout, body uniformly ciliated (Fig. 40C) .....

.... *Acropisthium* (p. 150)

30. Two transverse ciliary bands centrally placed each on either side of a waist-like furrow in body. Anterior ciliary band stiffened and cirri-like (Fig. 40D) .... *Mesodinium* (p. 158)

Two transverse ciliary bands which are not both centrally placed. When stiff, cirri-like cilia present, they are in the posterior band   **31**

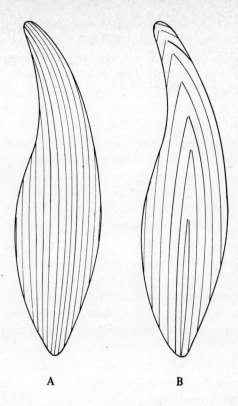

A   B

Fig. 38. Ciliary patterns on the upper surface of two gymnostome ciliates. A, *Litonotus*. B, *Amphileptus*.

31. Two transverse ciliary bands which are well separated from each other. No cirri-like cilia present (Fig. 40E) .... *Didinium* (p. 154)

Two transverse ciliary bands present which are closely set together in the anterior body half. Cirri may be present behind the posterior ciliary band. (Fig. 40F) .... *Askenasia* (p. 152)

32. With wart-like protuberances or processes containing trichocysts   **33**

Without warts or processes containing trichocysts **34**

33. Wart-like protuberances which are not extensible, confined to one body edge, never in apical or terminal region (Fig. 35C) ......

.... *Loxophyllum* (p. 174)

Extensible processes not confined to one body edge, but are situated in both apical and terminal regions .... *Lacerus* (p. 130)

34. Organisms spherical, forming chains .... *Sphaerobactrum* (p. 118)

When spherical never in chains **35**

35. Body barrel shaped, covered in plates with anterior and posterior spines .... *Coleps* (p. 103)

Never covered in plates, spines not present **36**

36. With distinct unciliated ridge containing trichocysts extends at least one-quarter length of body. Ridge may spiral in some genera. (Fig. 23) **37**

Without long unciliated ridge **42**

37. Body with two anterior horns (Fig. 23B) .... *Diceratula* (p. 127)

Body without anterior horns **38**

38. Unciliated ridge restricted to anterior half of body (Fig. 23E,F) **39**

Unciliated ridge extends into posterior half of body (Fig. 23A,C,D)   **40**

39. Body long and worm-like, hundreds of macronuclei present (Fig. 23F) .... *Myriokaryon* (p. 144)

Body flask shape, macronucleus moniliform, ectocommensal on crustacea (Fig. 23E) .... *Branchioecetes* (p. 142)

40. Unciliated ridge spirals around body and only just reaches terminal end (Fig. 23D) .... *Perispira* (p. 132)

Unciliated ridge travels down edge of body and curves back up around terminal end (Fig. 23A,C) **41**

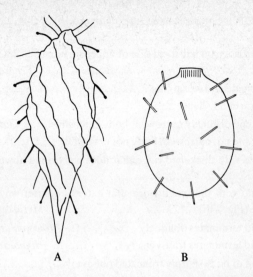

A   B

Fig. 39. Tentacle-like processes in two ciliates. A, *Dactylochlamys*. B, *Belonophrya*.

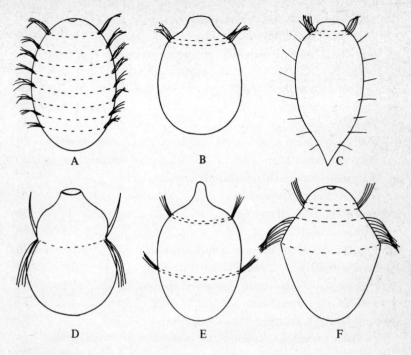

Fig. 40. Ciliary girdles of some gymnostome ciliates. Broken lines indicate boundaries of the girdles. A, *Dinophrya*. B, *Monodinium*. C, *Acropisthium*. D, *Mesodinium*. E, *Didinium*. F, *Askenasia*.

52. Apical end with distinct oblique unciliated ridge ..... *Spathidiodes* (p. 136)
Apical end oblique but without ridge .... *Peridionella* (p. 102)

53. Apical oral aperture with one or more lappets (short finger-like processes) **54**
Oral aperture without lappets **56**

54. With single lappet .... *Chilophrya* (p. 106)
With several lappets **55**

55. With caudal cilium .... *Plagiocampa* (p. 92)
Without caudal cilium .... *Spasmostoma* (p. 118)

56. Body sharply truncated transversely in apical region **57**
Body not sharply truncated **63**

57. With long caudal cilium/cilia **58**
Without caudal cilia **59**

58. Single caudal cilium, many long thick apical cilia .... *Peridion* (p. 100)
Several caudal cilia, apical cilia short .... *Crobylura* (p. 106)

59. With unciliated ridge around apical oral aperture **60**
Without ridge around oral aperture **61**

60. Body several times longer than width, trichocysts apical, with thick unciliated apical collar-like ridge (sometimes in mucilaginous lorica) .... *Enchelydium* (p. 128)
Body length about same as width, trichocysts in bunches all over body, thin unciliated ridge around wide oral aperture. Rare, only reported in Lake Baikal to date, planktonic .... *Spathidiosus* (p. 138)

61. Oral aperture apical, rounded **62**
Oral aperture apical, slit-like .... *Enchelys* (p. 109)

62. Oral aperture surrounded by spine-like cilia, aerobic in cave pools .... *Pseudoenchelys* (p. 168)
Oral aperture with cilia, anaerobic habitats .... *Pelatractus*(p. 84)

63. With caudal cilia/cilium **64**
Without caudal cilia **76**

64. With several caudal cilia **65**
With single caudal cilium **71**

77. Apical region ends in a dome-like, conical projection **78**

Apical region without projection **84**

78. Width of projection or collar one-quarter or less than maximum body width **79**

Width of projection or dome half body width **81**

79. Apical projection collar-like **80**

Apical projection conical shape .... *Lagynophrya* (p. 112)

80. Body ovoid with very small (one-sixth body width) apical collar ....
.... *Acaryophrya* (p. 78)

Body narrows posteriorly, medium sized (one-quarter body width) apical collar .... *Microchoanostoma* (p. 122)

81. Apical projection collar-like .... *Enchelydium* (p. 128)

Apical projection dome-like **82**

82. Macronucleus moniliform .... *Homalozoon* (p. 128)

Macronucleus in one or two parts **83**

83. With zoochlorellae .... *Microcardiosoma* (p. 122)

Without zoochlorellae .... *Enchelyodon* (p. 108)

84. Elongate oval with anterior semicircular indentation in one side (Fig. 25A) .... *Loxodes* (p. 76)

Without anterior lateral indentation **85**

85. Anterior oral aperture opens laterally, body shape irregular ......
.... *Amphibothrella* (p. 100)

Oral aperture apical, simple ovoid shape **86**

86. With conspicuous striations and ridges **87**

Without conspicuous ridges **88**

87. Ridges spiral in one direction, with lateral invagination in anterior body half .... *Placus* (p. 90)

Striations spiral in two directions at right angles to each other giving chequer board effect on surface, without lateral invagination .... *Baznosanuia* (p. 166)

88. Large apical depression in body leads to oral aperture (Fig. 22B)
.... *Bursellopsis* (p. 88)

Without apical depression although aperture is apically situated **89**

89. Oral aperture rounded **90**
    Oral aperture a slit **92**
90. With apical collar around oral aperture .... *Acaryophrya* (p. 78)
    Without collar around oral aperture **91**
91. Cytopharynx supported by conspicuous large double trichites ....
    .... *Prorodon* (p. 94)
    Cytopharynx supported by delicate basket of trichites ....
    .... *Holophrya* (p. 80)
92. Apical region truncated, flask-shaped body .... *Enchelys* (p. 109)
    Apical region rounded, ovoid body .... *Pseudoprorodon* (p. 98)

Sessile Gymnostomata

The single genus *Amphileptus* is the only sessile gymnostome, it attaches itself to peritrich colonies, upon which it preys, by means of a posterior thread (p. 172).

Order KARYORELECTIDA

Members of this order are, with one freshwater exception, all obligate marine interstitial forms. There is a dual nuclear apparatus as in other ciliates but here the macronucleus is diploid and does not divide. They tend to be fragile, highly thigmotactic and frequently elongate. *Loxodes* is the only freshwater example and it is laterally compressed with the somatic ciliature being concentrated on the right body side.

Loxodes Ehrenberg, 1830
(Fig. 41)

Description. Body outline elongated oval with anterior quarter pointed and strongly bent ventrally to form a concavity in which a slit-like oral aperture is situated. At the base of the slit lies a primitive pharyngeal basket of trichites. Body compressed laterally in the anterior region but less so in the posterior. Cilia mostly restricted to right (upper) surface, reduced on left (lower) surface to single kinety along each of the dorsal and ventral edges. Cilia usually in pairs forming up to 30 longitudinal parallel kineties. Many trichocysts present which give a brownish tint to the body. Several Muller's vesicles (vesicles containing spherical bodies of unknown function) may be found in the cell. Single contractile vacuole in posterior region. Macronucleus in one, two or more vesicular parts.

Key to species. Kahl (1930–35).

Morphology and ultrastructure. Dragesco (1966b, 1970), Penard (1917), de Puytorac and Njiné (1971).

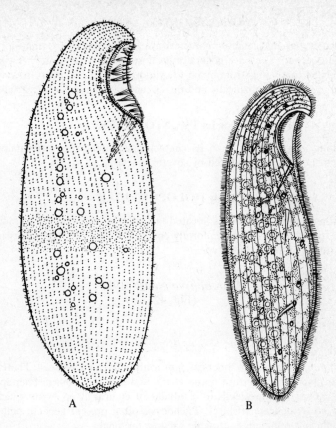

A   B

Fig. 41. *Loxodes*. A, After de Puytorac and Njiné (1971). B, After Roux (1899).

Order PROSTOMATIDA

This order has the general characteristics of the subclass Gymnostomata outlined earlier. They are generally an unspecialised group with a simple body shape that varies from ovoid to cylindrical. The oral aperture may be located apically or subapically and the circumoral ciliature is unspecialised.

Suborder PROSTOMATINA

Generally unspecialised in all its characteristics, they feed principally on bacteria. Loricas are produced by several species.

Family HOLOPHRYIDAE

The round oral aperture is located apically at the end of a radially symmetrical body which is uniformly covered in cilia. The cytopharyngeal apparatus is of the rhabdos type.

Acaryophrya André, 1915
(Fig. 42)

Balanophrya Kahl, 1930

Description. Body oval to spherical, uniformly ciliated all over cell. There is a circular apical oral aperture borne at the end of a short but distinct apical process. Oral aperture leads to a tubular or conical cytopharynx which is supported by delicate trichites. Trichocysts often present beneath pellicle. Macronucleus rounded, contractile vacuole terminal.

Most easily confused with *Holophrya* (p. 80) in which genus the oral aperture is flush with the apical surface and is not borne upon a short process.

Key to species. Kahl (1930–35).

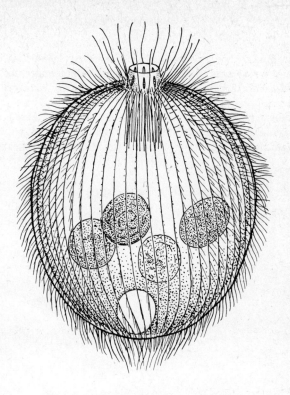

Fig. 42. *Acaryophrya* (after Kahl, 1930–35).

Holophrya Ehrenberg, 1833
(Fig. 43)

Description. Body shape ovoid to spherical, uniformly ciliated all over body except that in some species there may be a caudal tuft of longer cilia. There is a circular apical oral aperture which leads to a tubular or conical cytopharynx which is usually supported by a delicate basket of trichites. The oral aperture is a simple invagination and is flush with the apical surface, there is no apical collar-like process. Trichocysts often present all over body below the pellicle. Macronucleus ovoid. Contractile vacuole terminal. Feeds on other protozoa.

Key to species. Kahl (1930–35).

Descriptions. Dragesco, Iftode and Fryd-Versavel (1974).

Ultrastructure. de Puytorac (1965).

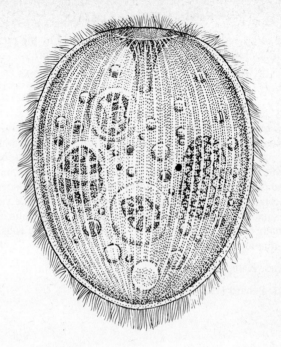

Fig. 43. *Holophrya* (after Mackinnon and Hawes, 1961).

Family METACYSTIDAE

They are generally unspecialised but are distinguished by the production of pseudochitinous loricas. One or more caudal cilia present.

Metacystis Cohn, 1866
(Fig. 44)

Description. Ovoid to cylindrical body, strongly striated transversely, usually tapering anteriorly. Apical circular oral aperture with about four rings of peribuccal cilia (see Bick, 1972) surrounding it. Posterior end broadly rounded with a peculiar protruding terminal clear vacuole. Somatic cilia uniform, usually with single (rarely more) caudal cilium which does not arise terminally. Macronucleus ovoid, centrally located. Contractile vacuole terminal with smaller lateral one. Constructs and lives in a membranous lorica which is usually open at one end only but occasionally at both ends. Found in anaerobic situations feeding on sulphur bacteria.

Key to species. Kahl (1930–35).

Descriptions. Penard (1922), Dietz (1964)

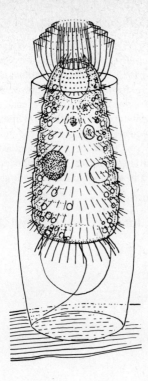

Fig. 44. *Metacystis* (composite from Bick, 1972; Dietz, 1964 and Kahl, 1930–35).

Pelatractus Kahl, 1930
(Fig. 45)

Description. Body shape elongate ovoid, greatest width about one-third of body length from anterior. Apical end truncated transversely. Somatic ciliation uniform, arising from bipolar kinetosomes which are somewhat irregularly arranged. Circular oral aperture apical, leading to wide permanent cytopharynx. The ciliature around the oral aperture (peribuccal cilia) is simple and consists of five rings of dense cilia, the most apical ring apparently being constructed of two or three cilia. No caudal tuft of cilia. Sometimes found in a thin membranous lorica. Macronucleus rounded, contractile vacuole terminal. Found in anaerobic muds feeding on sulphur bacteria.

Key to species. Kahl (1930–35).

Descriptions. Dragesco, Iftode and Fryd-Versavel (1974), Penard (1922).

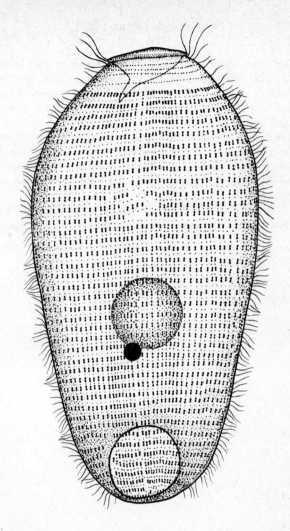

Fig. 45. *Pelatractus* (composite from Dragesco, Iftode and Fryd-Versavel, 1974).

Vasicola Tatem, 1869
(Fig. 46)

Description. Body shape irregularly ovoid, sometimes with broad anterior end, sometimes widest in middle of body. Circular oral aperture apically located leading to a conical cytopharynx that is supported by trichites. Somatic ciliation uniform arising from bipolar kinetosomes, with a group of half a dozen long caudal cilia. There are three groups of cilia encircling the mouth, that nearest the oral aperture consists of a single row, the next is a double row and the ring nearest the somatic cilia consists of four closely set ciliary rows. Lives in a thin, transparent membranous lorica which has a single narrow aperture. Inhabits anaerobic habitats where it feeds upon sulphur bacteria.

Most easily confused with *Pelatractus* (p. 84) which does not have caudal cilia and has a different arrangement of peribuccal cilia.

Descriptions. Dragesco, Iftode and Fryd-Versavel (1974), Grolière (1977), Kahl (1930–35), Penard (1922).

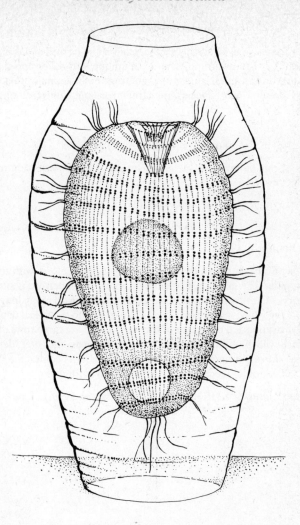

Fig. 46. *Vasicola* (composite from Dragesco, Iftode and Fryd-Versavel, 1974).

Suborder PRORODONTINA

The oral aperture may be apical or subapical, round or oval and is sometimes at the base of a shallow atrium. There is a unique 'brosse' or dorsal brush of cilia on the dorsal surface near the anterior pole. Somatic toxicysts are commonly present. Most species are carnivorous but some feed on bacteria and algae.

Family PRORODONTIDAE

Members of this family have the characteristics of the above suborder.

Bursellopsis Corliss, 1960
(Fig. 47)

Bursella Schmidt, 1921

Description. Body shape either ovoid or irregularly pyriform with truncated anterior region. The large oval or circular apical oral aperture is situated at the base of a wide shallow depression. Body covered with many longitudinal kineties of cilia which curve over the apical region into the apical depression. Cytopharynx supported by basket of trichites. There are three rows of double kinetosomes which either straddle or run towards the oral depression forming a dorsal 'brush'. Macronucleus rounded, contractile vacuole posterior. Planktonic, feeds upon flagellates and rotifers.

Descriptions. Dragesco, Iftode and Fryd-Versavel (1974), Fauré-Fremiet (1924).

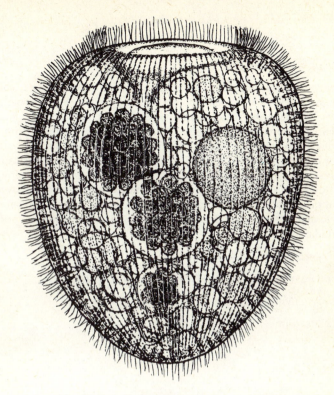

Fig. 47. *Bursellopsis* (after Fauré-Fremiet, 1924).

Placus Cohn, 1866
(Fig. 48)

Thoracophrya Kahl, 1926

Description. Body outline shape oval, small (30–70 μm long), slightly compressed. Pellicle with conspicuously spiral ridges with distinct spots in the furrows between the ridges. Oral aperture oval and apically located, cytopharynx supported by trichites. Somatic ciliature uniform, with cilia emerging from spiral meridians of paired kinetosomes. There is a fibril (shown in protargol preparations) on the left of each pair of kinetosomes. The dorsal 'brush' is composed of a double parallel row of kinetosomes which arises adjacent to the oral aperture and terminates almost halfway down the body at the site of a lateral invagination of unknown function. Macronucleus ovoid to elongate, centrally located. Contractile vacuole either posterior or posterio-lateral.

Key to species. Kahl (1930–35).

Descriptions. Foissner (1972), Fryd-Versavel *et al.* (1976).

Stomatogenesis. Riordan and Small (1975).

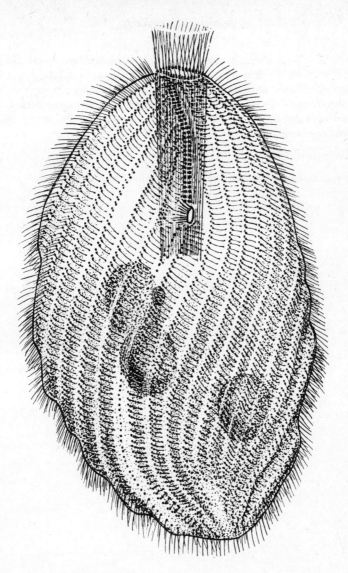

Fig. 48. *Placus* (after Fryd-Versavel, Iftode and Dragesco, 1976).

Plagiocampa Schewiakoff, 1892
(Fig. 49)

Description. Body shape ovoid to cylindrical. Oral aperture apical, slit-like with a thickened ridge along right edge bearing several long finger-like protoplasmic processes which contain trichocysts. Cytopharynx supported by trichites. Somatic ciliation uniform and complete, kinetosomes paired. Some species with long caudal cilium. Macronucleus oval, centrally located. Single contractile vacuole, posterio-laterally situated.

Key to species. Kahl (1930–35).

Description. Fauré-Fremiet and André (1965b), Foissner (1978b).

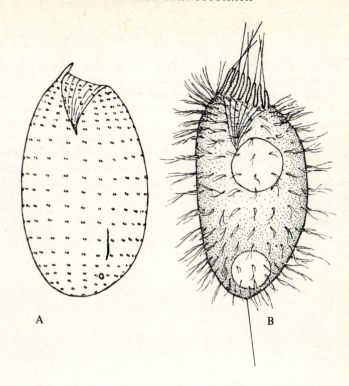

A   B

Fig. 49. *Plagiocampa*. A, Silver impregnated specimen. B, Whole animal (after Fauré-Fremiet and André, 1965b).

Prorodon Ehrenberg, 1833
(Fig. 50)

Description. Body shape ovoid to cylindrical, in some species the body narrows posteriorly. Oral aperture circular, apical, leading to a cytopharynx which is supported by large, distinct double trichites. Somatic ciliation uniform, in longitudinal meridians whose kinetosomes may be paired. The meridians with one or two exceptions reach the apical oral aperture. There are three parallel double rows of kinetosomes above the short meridians and these comprise the basal bodies of the dorsal 'brush'. A few species have a tuft of caudal cilia. Macronucleus rounded to elongate. One or more contractile vacuoles in posterior region. Some species feed upon protozoa and algae, others are histophagous.

Key to species. Kahl (1930–35).

Descriptions. Dragesco (1966b, 1970), Grolière (1977), Jordan (1974), de Puytorac and Savoie (1968).

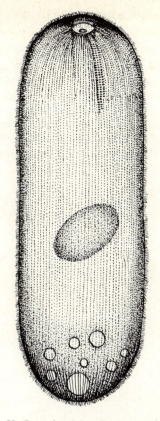

Fig. 50. *Prorodon* (after Grolière, 1977).

Urotricha Claparède and Lachmann, 1859
(Fig. 51)

Description. Body shape ovoid to spherical. Oral aperture apical, cytopharynx supported by trichites. Somatic cilia uniform but longitudinal kineties only extend part way down the cell so that the posterior body quarter is free from cilia with the exception of one or more caudal cilia. There are one or two circlets of double cilia surrounding the oral aperture (peribuccal cilia) and a dorsal brush consisting usually of three or four short lines of paired cilia. Macronucleus spherical to ovoid, usually centrally situated. One or more contractile vacuoles in the posterior cilia-free zone. Some species with trichocysts.

Key to species. Kahl (1930–35).

Descriptions. Dragesco *et al.* (1974), Grolière (1977).

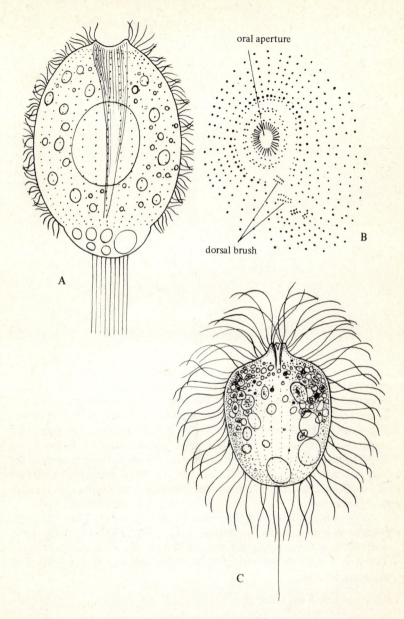

Fig. 51. *Urotricha.* A, Whole animal. B, Oral area after silver impregnation (both after Dragesco, Iftode and Fryd-Versavel, 1974). C, Whole animal (from life).

Pseudoprorodon Kahl, 1930

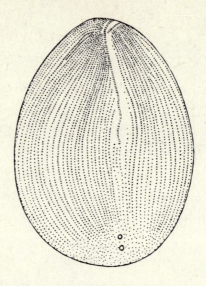

Fig. 52. *Pseudoprorodon* (after Grolière, 1977).

Description. Body outline shape usually elongate but sometimes oval, often laterally flattened. Oral aperture is an apically situated long slit which is supported by many fine trichites. The oral slit is surrounded by a row of double kinetosomes (peribuccal ciliature) which extends past the slit on one edge down to the middle of the body. The dorsal 'brush' is represented by a single short row of closely packed kinetosomes which lies between and parallel with the longitudinal somatic kineties in the anterior quarter of the body although unlike the somatic kineties, the dorsal brush does not reach the peribuccal kinety. Macronucleus elongate, ribbon-like. Contractile vacuole terminal.

Most easily confused with *Prorodon* (p. 94) which has fewer but larger (double) trichites and a rounded apical oral aperture. Furthermore the dorsal brush of *Prorodon* is composed of three kineties of paired kinetosomes.

Generic name was erected by Kahl (1930) although he attributes it to Blochmann (1886).

Key to species. Kahl (1930–35).

Description. Grolière (1977).

Rhagadostoma Kahl, 1926

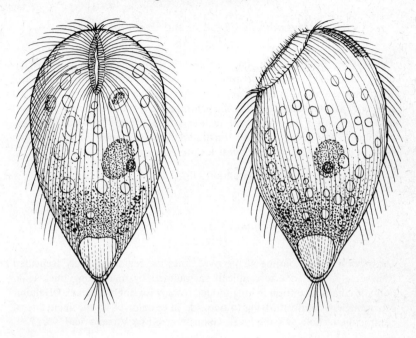

Fig. 53. *Rhagadostoma*. A. Ventral surface. B, Lateral view (after Kahl, 1926).

Description. Body outline shape pyriform, anterior region broader than posterior. Apical oral aperture slit-like borne on apical ridge, cytopharynx supported by distinct, double trichites. Somatic ciliation poorly described, uniform in longitudinal meridians, three rows of cilia form dorsal brush. In one species the somatic cilia do not cover the posterior part (as in *Urotricha* p. 96) but there is always a tuft of caudal cilia. Macronucleus spherical to ovoid, centrally located. Single terminal contractile vacuole. Sapropelic.

Description. Kahl (1926, 1930–35).

INCERTAE SEDIS PRORODONTIDAE

Amphibothrella Grandori and Grandori, 1934
(Fig. 54)

Description. Body outline shape irregularly oval, slightly compressed laterally. Dorsal side convex, ventral side concave but with median bulge. Oral aperture large and rounded at anterior end of cell on ventral edge. There is a concavity at the posterior end of the ventral edge (for attachment purposes?) associated with a large adjacent vacuole which is not contractile. Somatic ciliation in the form of many longitudinal kineties with an apical patch without cilia. Several Muller's vesicles present.

Descriptions. Grandori and Grandori (1934, 1935).

Peridion Vuxanovici, 1962
(Fig. 55)

Description. Body outline shape oval, anterior region severely truncated transversely. Oral aperture apical, surrounded by many long thick cilia. Somatic ciliation uniform in longitudinal rows, with single long caudal cilium. Macronucleus reniform, lying to one side in equator of body, single lateral contractile vacuole at same level. Genus erected by Vuxanovici (1962a).

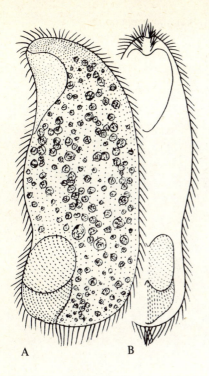

Fig. 54. *Amphibothrella*. A, Lateral aspect. B, Ventral aspect (after Grandori and Grandori, 1934).

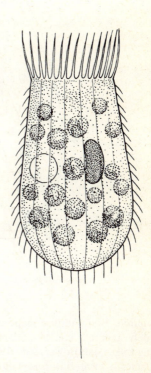

Fig. 55. *Peridion* (after Vuxanovici, 1962a).

Peridionella Vuxanovici, 1963

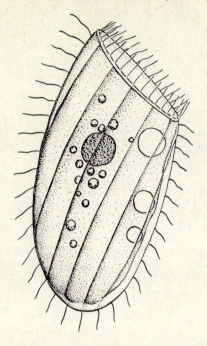

Fig. 56. *Peridionella* (after Vuxanovici, 1963).

Description. Body outline shape oval with anterior end severely truncated obliquely. Laterally compressed, right (usually uppermost) side with distinct oblique ridges, left (lower) side flattened but slightly concave. Oral aperture oval, apically located with row of cilia along right edge only. Macronucleus spherical, centrally located. Three contractile vacuoles along ventral edge with long serving canal which extends down ventral edge and around terminal region of cell.

Family COLEPIDAE

Members of the family have the same general features of the suborder but they have a characteristic barrel-shaped body that is covered in calcified armoured plates often bearing small lateral teeth and prominent anterior and posterior spines. Caudal cilia are commonly present.

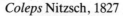

Coleps Nitzsch, 1827

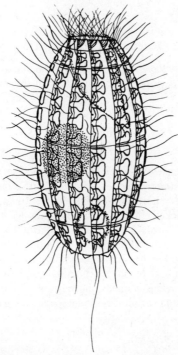

Fig. 57. *Coleps* (after Wilbert and Schmall, 1976).

Description. Body barrel-shaped, covered in regularly arranged prominent ectoplasmic plates that are composed of amorphous calcium carbonate. Each end of the cell is rounded or slightly flattened, never pointed but often with tooth-like projections from the plates. Oral aperture circular, apically situated. Somatic ciliature uniform in regular longitudinal kineties along the striations in the plates. One or more caudal cilia present but easily overlooked. Anterior end with two rings of peribuccal cilia and a dorsal brush. Macronucleus ovoid, contractile vacuole posterior.

Keys to species. Kahl (1930–35), Noland (1925).

Descriptions. Fauré-Fremiet and Hamard (1940), Gieman (1931), Kahl (1930), Noland (1925), Wilbert and Schmall (1976).

Order HAPTORIDA

The oral aperture is apically or subapically located and may be oval or slit-like such that it is sometimes not permanently open. The cytopharynx is reversible in some species and there is often a field of clavate 'sensory' cilia present. The toxicysts are usually localised, typically in or near the oral area. The rhabdos-like cytopharyngeal basket is complex. These ciliates are rapacious carnivores, some being equipped with a proboscis and some with non-suctorial tentacles.

Family ENCHELYIDAE

Members of this family have the characters of the above order. In several of the species the oral aperture is located distally at the end of a long flexible neck region.

Chaenea Quennerstedt, 1867
(Fig. 58)

Description. Body shape always elongate, rounded in cross-section. Body highly contractile. Posterior may be rounded or with a narrowing tail region. Anterior always with a short, non-extensible snout-like region which bears many more closely packed cilia (and may also be longer) than the rest of the body. Oral aperture oval, located apically. Body covered in uniform longitudinally arranged ciliary rows which commonly spiral slightly. Three of these rows have paired kinetosomes in the anterior region close to the snout, these represent the dorsal brush. Trichocysts common, particularly in the snout region. Macronucleus in one or more rounded parts. There is always a large terminal contractile vacuole and some species have extra lateral rows of vacuoles.

Key to species. Kahl (1930–35).

Descriptions. Fauré-Fremiet and Ganier (1970), Fryd-Versavel *et al.* (1976).

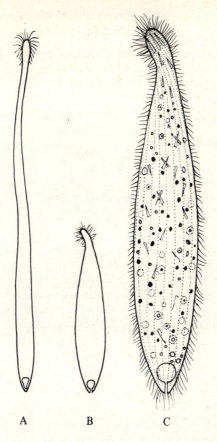

A   B   C

Fig. 58. *Chaenea.* A, Body extended. B, Body contracted. C, Detail (after Dragesco, 1966b).

Chilophrya Kahl, 1930
(Fig. 59)

Description. Body shape ovoid, apical oral aperture with a lip-like projection on one side folded over aperture. Ciliation uniform in longitudinal rows. One species found in inland salt lakes contains zoochlorellae. Macronucleus ovoid, centrally located. Contractile vacuole terminal.

Key to species. Kahl (1930–35).

Description. Edmondson (1920).

Crobylura André, 1914
(Fig. 60)

Description. Body shape elongated oval with severely transversely truncated anterior end. Body rather flexible and metabolic. Oral aperture apical, without permanent cytopharynx. Body covered in short cilia, including anterior border. Posterior region with bunch of prominent caudal cilia. Many trichocysts arranged in longitudinal rows in anterior body half. Macronucleus spherical in anterior body half. Contractile vacuole in posterior region. Planktonic. Single record, Lake Leman, Switzerland.

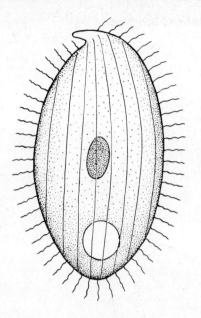

Fig. 59. *Chilophrya* (after Edmondson, 1920).

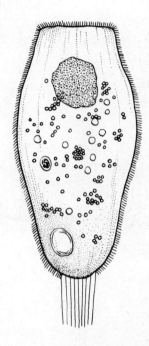

Fig. 60. *Crobylura* (after André, 1914).

Enchelyodon Claparède and Lachmann, 1859

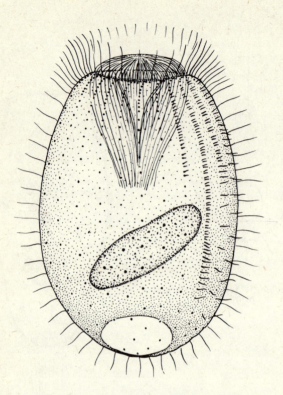

Fig. 61. *Enchelyodon* (after Grolière, 1977).

Description. Body outline shape rounded, flask-like, elongate or oval, rounded in cross-section. Anterior domed or with very short, but distinct, snout-like region which is usually armed with trichocysts and often surrounded by long cilia. Oral aperture at end of snout region. Cytopharynx supported by trichites. Somatic ciliation uniform and complete with dorsal brush formed from anterior ends of one to four kineties whose ciliary bases are often paired. Usually single macronucleus, rounded to elongate, rarely the macronucleus is in two parts. Contractile vacuole posterior.

Key to species. Kahl (1930–35).

Descriptions. Dragesco (1970), Grolière (1977), Vuxanovici (1963).

Enchelys Müller, 1773

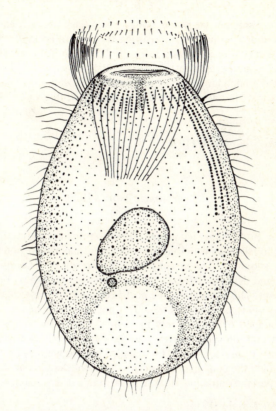

Fig. 62. *Enchelys* (after Dragesco, Iftode and Fryd-Versavel, 1974).

Description. Body outline shape flask-like to oval whose anterior end is always truncated transversely. Oral aperture apical, slit-like, whose edge is often slightly concave but never domed. Cytopharynx supported by trichites. Cilia around edge of oral aperture often long and always more closely packed than on the rest of the body. Somatic ciliation complete and uniform except for three kineties whose anterior ends are very closely packed. These represent the dorsal brush. Macronucleus varies from spherical to elongate and ribbon-like. Contractile vacuole usually single and terminal but rarely with large terminal vacuole plus several smaller vacuoles.

Key to species. Kahl (1930–35).

Descriptions. Dragesco *et al.* (1974), Fauré-Fremiet (1944b), Vuxanovici (1963).

Ileonema Stokes, 1884
(Fig. 63)

Description. Body outline shape elongate and flask-like, flattened laterally. Broadly rounded posteriorly, narrowing towards anterior. Rather contractile. Apical region bears a long retractile flagellum-like organelle which in one species is borne upon a thicker, twisted retractile rod. Body ciliation uniform and complete. Cytopharynx armed with delicate trichites. Macronucleus rounded in one or two parts. Single contractile vacuole terminal.

Key to species. Kahl (1930–35).

Description. Schewiakoff (1893).

Lacrymaria Bory, 1826
(Fig. 64)

Description. Body shape changeable due to contractility, varies from ovoid to cylindrical to flask-like with long anterior extensible neck region upon which there is a snout-like region consisting of two zones. The apical zone is unciliated and demarks the area which carries the temporary oral aperture when present. The zone nearer the neck is ciliated with long cilia which arise from many short oblique kineties. The posterior end of the cell may be broadly rounded or may narrow to a blunt point. The cell is completely covered by uniform cilia (shorter than those on the snout) which arise from longitudinal or spiral kineties. The somatic kinetosomes may be paired at the anterior end of the body in some species. Macronucleus in one, two or rarely in several parts. Contractile vacuole in posterior body region.

Key to species. Kahl (1930–35).

Descriptions and structure. Bohatier (1971), Buitkamp and Wilbert (1974), Ehrenberg (1830), Fauré-Fremiet (1924), Hovasse (1938), Kahl (1926), Kink (1972).

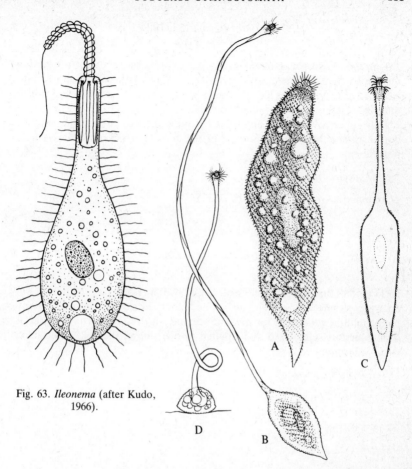

Fig. 63. *Ileonema* (after Kudo, 1966).

Fig. 64. *Lacrymaria*. A, Animal fully contracted. B, Fully extended. C, Usual swimming position. D, Settled on substratum with neck fully extended (all drawn from life).

Lagynophrya Kahl, 1927
(Fig. 65)

Description. Body shape ovoid, cylindrical to pyriform, often slightly bent to observer's right (animal's ventral surface). Apically there is a short, non-ciliated, retractile, cone-like proboscis containing trichocysts below which there is a basket of fine, paired trichites. Cell covered in cilia arranged in slightly spiral rows. There are a few clavate (clubbed) cilia on the anterior dorsal surface. Macronucleus ovoid in one or two parts. Contractile vacuole posterior.

Key to species. Kahl (1930–35).

Descriptions. Grain (1970), Kahl (1927).

Longitricha Gajewskaja, 1933
(Fig. 66)

Description. Body shape spherical, slightly compressed pole to pole. Oral aperture situated at apical end of body but difficult to observe. Body pigmented brown except for posterior region which is completely devoid of cilia. Macronucleus ovoid and situated to one side of the cell. Contractile vacuole terminal.

Description. Gajewskaja (1933).

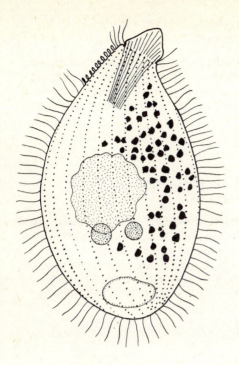

Fig. 65. *Lagynophrya* (after Grain, 1970).

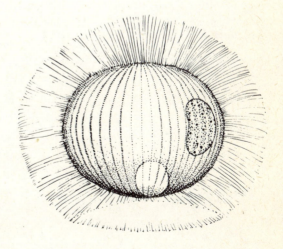

Fig. 66. *Longitricha* (after Gajewskaja, 1933).

Microregma Kahl, 1930
(Fig. 67)

Description. Body shape ovoid, small (40–50 µm long). Oral aperture apical, slit-like surrounded by long cilia. Somatic cilia uniform, covering entire body. Freshwater species have a caudal cilium. Macronucleus spherical, central. Contractile vacuole laterally located in posterior body region.

Key to species. Kahl (1930–35).

Pithothorax Kahl, 1926
(Fig. 68)

Description. Body shape ovoid to cylindrical. Oral aperture rounded, apical and surrounded by long cilia. Pellicle rigid, longitudinally ribbed in some species. Somatic ciliature reduced in central region, cilia present on anterior and posterior quarters of body only. Caudal cilium present. Macronucleus spherical in centre or anterior half of cell. Single contractile vacuole in posterior region.

Key to species. Kahl (1930–35).

Description. Kahl (1926).

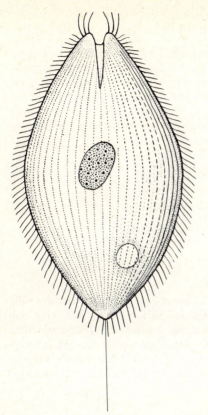

Fig. 67. *Microregma* (after Kudo, 1966).

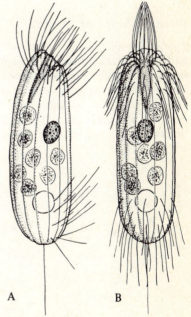

A   B

Fig. 68. *Pithothorax*. A, Lateral view. B, Ventral view (after Kahl, 1926).

Rhopalophrya Kahl, 1926
(Fig. 69)

Description. Body outline shape elongate, flask-like, sometimes slightly asymmetrical, pellicle with widely separated longitudinal ribs and furrows. Some species rounded in cross-section, others somewhat compressed. Oral aperture apical at end of short neck or snout region. Uniform cilia cover the body, those near the oral aperture longer than rest. Usually one (rarely two) spherical macronucleus, centrally located. Contractile vacuole terminal.

Key to species. Kahl (1930–35).

Descriptions. Kahl (1926), Vuxanovici (1962a).

Schewiakoffia Corliss, 1960
(Fig. 70)

Maupasia Schewiakoff, 1892

Description. Body shape approximately ovoid but rather asymmetrical with flattened ventral surface and convex dorsal surface. Body metabolic and slightly contractile. Oral aperture in anterior quarter of body opening from ventral surface. Anterior somatic cilia denser than posterior cilia. Caudal cilium present. Macronucleus oval, centrally located. Contractile vacuole posterior with permanent pore opening onto ventral surface. Rare, description based on single record in Russia.

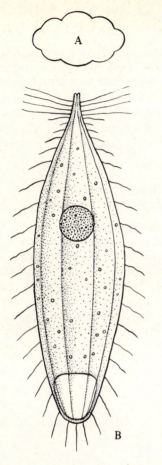

Fig. 69. *Rhopalophrya*. A, Transverse section. B, Detail (after Vuxanovici, 1962a).

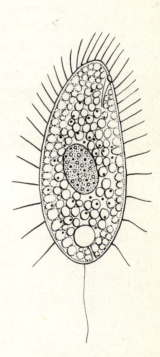

Fig. 70. *Schewiakoffia* (after Schewiakoff, 1893).

Spasmostoma Kahl, 1927
(Fig. 71)

Description. Body shape ovoid. Oral aperture apical, oval, surrounded by about 20 finger-like flaps or lappets. Cytopharynx supported by trichites. Body entirely covered by uniform cilia arranged in longitudinal kineties. Caudal cilium absent. Feeds on flagellates which colour food vacuoles bright green. Macronucleus spherical, centrally located. Single contractile vacuole.

Description. Kahl (1927).

Sphaerobactrum Schmidt, 1920
(Fig. 72)

Description. Body shape spherical, usually found in chains of four individuals (length of chain 600 μm). Body covered in many short cilia. Oral aperture difficult to distinguish. With zoochlorellae in cytoplasm. Without a contractile vacuole but periphery of cell with many vacuoles. Macronucleus rounded, centrally located. Planktonic. Description based on single report in Germany.

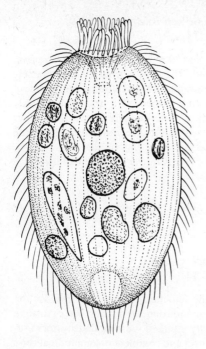

Fig. 71. *Spasmostoma* (after Kahl, 1927).

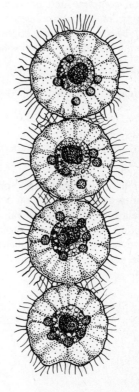

Fig. 72. *Sphaerobactrum* (after Schmidt, 1920).

Trachelophyllum Claparède and Lachmann, 1859
(Fig. 73)

Description. Body shape elongate, flask-like with anterior neck region. Highly flexible body and neck, flattened. The neck is truncated transversely and carries a short snout-like tip. Oral aperture apically located on snout leading to narrow rounded cytopharynx armed with trichocysts. Body ciliation complete and uniform. Cilia at anterior end longer than other somatic cilia. There is a line of short bristle-like cilia, which represents the dorsal brush, down one edge of the extreme end of the neck region. Macronucleus always in two spherical parts. Contractile vacuole terminal.

Key to species. Kahl (1930–35).

Descriptions. Dragesco (1966a), Kahl (1926), Vuxanovici (1959b).

Urochaenia Savi, 1913
(Fig. 74)

Description. Body elongate, contractile and flexible, oval in cross-section with anterior neck-like region. Posterior end rounded bearing two cirrus-like organelles. Oral aperture oval, apically situated, surrounded by forward-projecting long cilia. Somatic cilia uniform and complete in longitudinal rows that do not spiral. Macronucleus rounded, centrally placed. Contractile vacuole terminal. Single record in stagnant crater lake in Italy.

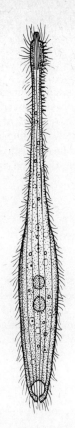

Fig. 73. *Trachelophyllum* (after Dragesco, 1966a).

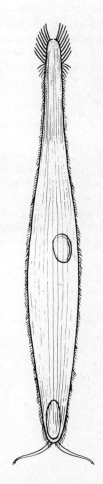

Fig. 74. *Urochaenia* (after Savi, 1913).

INCERTAE SEDIS ENCHELYIDAE

Microcardiosoma Vuxanovici, 1963
(Fig. 75)

Description. Body outline shape heart-like to oval, rounded in cross-section with regular longitudinal ribs. Oral aperture apical, located at end of distinct but short anterior snout which is surrounded by a group of long cilia projecting forwards. Somatic ciliation uniform, entirely covering body which is packed with numerous green zoochlorellae. Macronucleus spherical, located in centre of cell. Contractile vacuole not seen. Based on single report in Romania by Vuxanovici (1963).

Microchoanostoma Vuxanovici, 1963
(Fig. 76)

Description. Outline shape heart-like, laterally flattened. There is a short but mobile anterior snout supported by trichites upon which the apical oral aperture is situated. Body covered in longitudinal striations and is entirely and uniformly ciliated. Central macronucleus is reniform. Large terminal contractile vacuole. Based on report in Romania by Vuxanovici (1963) on a single example found in a sapropelic sample.

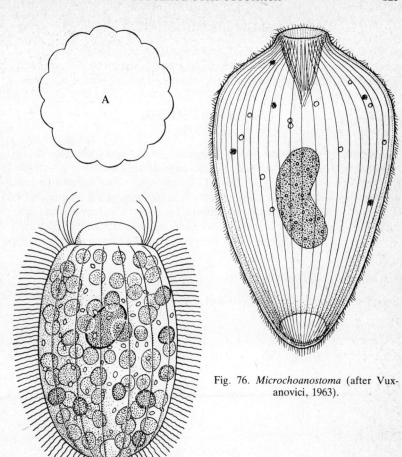

Fig. 76. *Microchoanostoma* (after Vuxanovici, 1963).

Fig. 75. *Microcardiosoma.* A, Transverse section. B, Lateral aspect (after Vuxanovici, 1963).

Family SPATHIDIIDAE

In this family the oral aperture is slit-like and is generally located apically on a non-ciliated ridge of the body which facilitates the ingestion of large prey. The body is often flask or sack shaped, is flattened and is truncate anteriorly.

Bryophyllum Kahl, 1931
(Fig. 77)

Description. Body outline shape irregularly oval (rarely elongate as in *Amphileptus*), laterally flattened. Ventral edge strongly convex on which the long slit-like oral aperture is situated. Dorsal edge either concave or less convex than ventral edge whose border is clear forming a C-shaped band lined with trichocysts. Ciliation uniform on both right (upper) and left (lower) surfaces. Single contractile vacuole in posterior quarter of body. Macronucleus either elongate and ribbon-like, moniliform or rarely single oval mass.

Key to species. Kahl (1930–35).

Descriptions of species. Fryd-Versavel *et al.* (1976), Gelei (1933).

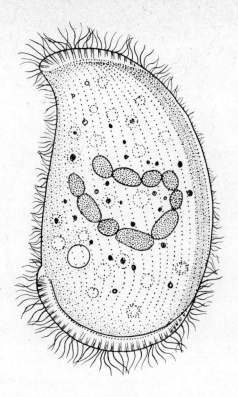

Fig. 77. *Bryophyllum* Whole animal. (after Fryd-Versavel, Iftode and Bragesco, 1976).

Cranotheridium Schewiakoff, 1892

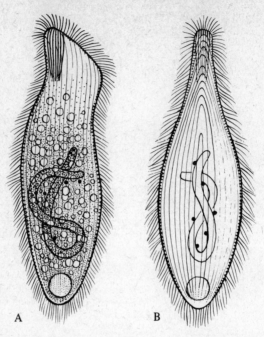

Fig. 78. *Cranotheridium*. A, Lateral aspect. B, Ventral aspect (after Schewiakoff, 1892).

Description. Body elongate, rounded in cross-section, posterior bluntly rounded. Anterior region of body characteristically terminates obliquely to the longitudinal body axis with a narrowed region behind the apex. Uniformly ciliated in parallel rows which meet anteriorly along the apical edge which may or may not bear an unciliated ridge. The oral aperture is rounded, apically situated and supported by trichites. Large terminal contractile vacuole with or without some small vacuoles on the ventral surface. Macronucleus either elongate, ribbon-like and twisted, or ovoid.

Most easily confused with the genus *Spathidium* (p. 138) which has a slit-like oral aperture unsupported by trichites whereas *Cranotheridium* has a rounded, supported oral aperture.

Description of species. Vuxanovici (1962b).

Diceratula Corliss, 1960

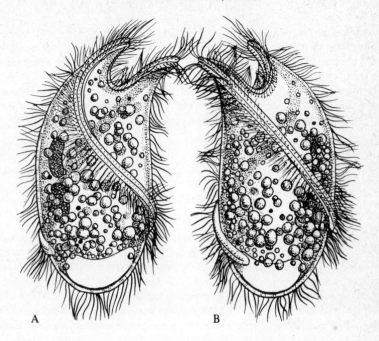

Fig. 79. *Diceratula*. A, Lateral aspect. B, Lateral aspect (after Kahl, 1930–35).

Diceras Eberhard, 1862

Description. Body outline shape oval except for the anterior end which is divided into two obliquely directed horn-like projections. The apical concavity formed between the horns is lined by an unciliated ridge in which the slit-like oral aperture is located. As in *Perispira* (p. 132), the ridge continues past the apical region and spirals down around the body. Ciliation uniform. Macronucleus in two sausage-shaped parts. Large single terminal contractile vacuole present.

Description of species. Kahl (1930).

Enchelydium Kahl, 1930
(Fig. 80)

Description. Body shape elongate to ovoid, rounded in cross-section, slightly contractile. Anterior end truncated transversely with definite swollen apical ridge which encircles circular or elongate oral aperture. The aperture is armed with trichocysts and appears to open when swimming. Ciliation uniform with row of dorsal bristles. One species constructs a gelatinous lorica in which it lives. One or two macronuclei present which may be ovoid to elongate. Single posterior contractile vacuole.

Description of species. Kahl (1930).

Homalozoon Stokes, 1890
(Fig. 81)

Description. Body shape elongate (150–1500 μm long), worm-like, flattened laterally, contractile. Posterior with tail-like region or bluntly rounded. Anterior end blunt, slanting slightly posterior from right to left. Oral aperture a slit along the apical edge which consists of a well-defined thickened ridge. Oral region supported by trichites, ridge unciliated but cilia are situated just behind it. The left (upper) surface has three kineties but is devoid of cilia, in some species there is a distinctive longitudinal ridge along its mid-line. The right (lower) surface is ciliated with several (up to about 20) longitudinal kineties. Trichocysts distributed throughout body. Macronucleus usually moniliform, usually along ventral edge. Several contractile vacuoles in line along dorsal edge.

Key to species. Kahl (1930–35).

Descriptions and morphology. Diller (1964), Dragesco (1966b), Fryd-Versavel *et al.* (1976), Girgla (1971), Weinreb (1955a,b).

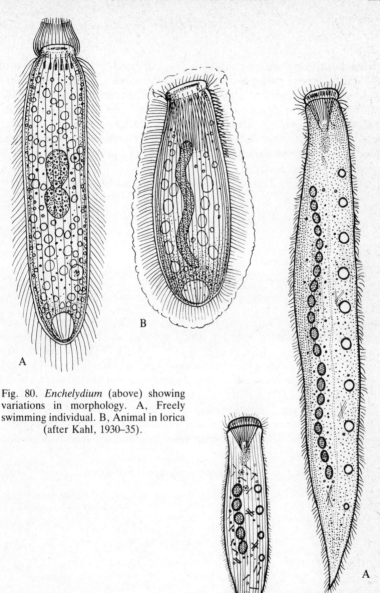

Fig. 80. *Enchelydium* (above) showing variations in morphology. A, Freely swimming individual. B, Animal in lorica (after Kahl, 1930–35).

Fig. 81. *Homalozoon* (right). A, Typical elongate form. B, Shorter species (after Dragesco, 1966b).

Lacerus Jankowski, 1967
(Fig. 82)

Description. Body shape approximately elongated oval in outline with tail-like process in posterior and with truncated anterior region. The apex slants slightly from left to right and bears the slit-like oral aperture. The animal has several extensible processes on the body edges which contain trichocysts. There is one process on either side of the apex, several down the ventral edge, one forms the tail and there are two on the posterior part of the dorsal edge. The macronucleus is sausage-shaped and the single contractile vacuole lies in the posterior body third.

Genus erected by Jankowski (1967c) for *Legendrea pespelicani* Penard, 1922.

Description of species. Penard (1922).

Legendrea Fauré-Fremiet, 1908
(Fig. 83)

Description. Body shape irregularly oval in outline with broadly round posterior and truncated anterior. Body narrows anteriorly, apical region slants slightly posteriorly from right to left. Unciliated ridge bearing slit-like oral aperture borne upon the slanted apex. Oral aperture supported by very large basket of trichites which occupies the anterior third of the body. On the left (upper) surface there are about 20 finger-like structures of variable length, each is slightly dilated at its distal tip in which trichocysts are located. These processes are plastic but not capable of controlled movement. Macronucleus elongate. Contractile vacuole either on ventral edge with canals or large and terminal.

Description of species. Fauré-Fremiet (1908).

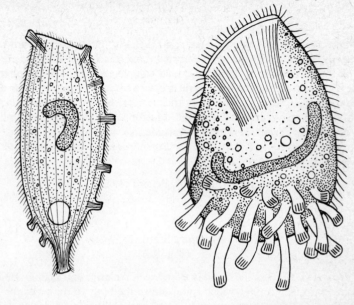

Fig. 82. *Lacerus* (after Kahl, 1930–35).   Fig. 83. *Legendrea* (after Fauré-Fremiet, 1908).

Penardiella Kahl, 1930
(Fig. 84)

Description. Body outline shape oval, laterally flattened. Anterior region slightly truncated. Prominent unciliated ridge travels around complete body edges (but sometimes does not extend completely up the dorsal edge) but never spirals around the body. Oral aperture situated towards dorsal edge on apical part of ridge, supported by trichites. Trichocysts line the body ridge. Cilia uniformly distributed over cell. Macronucleus elongate, ribbon-like or ovoid. Single terminal contractile vacuole.

Genus erected by Kahl (1930) for *Legendrea interrupta* Penard, 1922 and *Legendrea crassa* Penard, 1922.

Descriptions of species. Kahl (1930), Penard (1922).

Perispira Stein, 1859
(Fig. 85)

Description. Body shape ovoid to cylindrical. Anterior end slightly truncated bearing unciliated ridge which spirals down the body to the posterior. Oral aperture is slit-like, borne upon the apical part of the ridge. Ciliation uniform either in spiral or longitudinal kineties. Macronucleus elongate to ovoid. Single terminal contractile vacuole.

Genus erected by Stein (1859b).

Descriptions of species. Dewey and Kidder (1940), Kahl (1930).

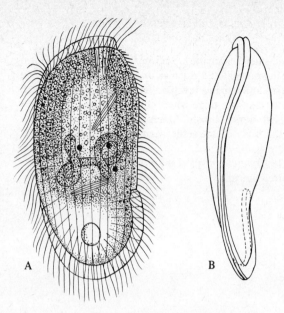

Fig. 84. *Penardiella*. A, Lateral view.
B, Ventral view showing unciliated
ridge (after Kahl, 1930–35).

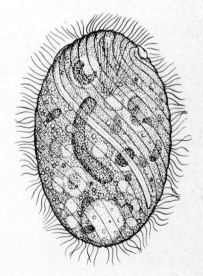

Fig. 85. *Perispira* with spiralling unciliated ridge (after Dewey and Kidder,
1940).

Protospathidium Dragesco and Dragesco–Kerneis, 1979
(Fig. 86)

Description. Body outline shape elongate, posterior bluntly rounded, apical region ending obliquely in an unciliated ridge as in *Spathidium* but here the ridge is very short. Ciliation uniform in a few parallel longitudinal rows which bend obliquely apically to meet the ridge where they are paired for a short distance. There is a short dorsal brush. Macronucleus in several parts distributed throughout the body. Contractile vacuole single and terminal. Feeds on flagellates.

Most easily mistaken for *Spathidium* (p. 138) in which genus the kineties neither bend nor are paired apically.

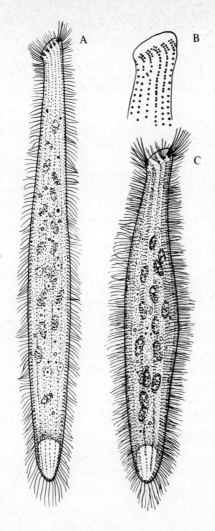

Fig. 86. *Protospathidium*. A, Whole animal. B, Silver impregnated specimen showing paired bent apical kineties. C, Whole animal (after Dragesco and Dragesco-Kerneis, 1979).

Spathidiodes Kahl, 1926
(Fig. 87)

Spathidiella Kahl, 1930

Description. Body outline shape oval to elongate, usually with rounded posterior but one species with pointed posterior. Body rigid not contractile. Anterior region always truncated obliquely to major body axis, bearing an unciliated apical ridge which is narrow towards the ventral edge and widens (sometimes to form a beak-like structure) towards the dorsal edge. Oral aperture slit-like, borne on apical ridge, when trichocysts are present they line the complete apical ridge without warts. Body ciliation uniform with dorsal brush of bristles. Single ovoid macronucleus present. Single posterior contractile vacuole.

Key to species. Kahl (1930–35).

Descriptions of species. Kahl (1926, 1930).

Spathidioides Brodsky, 1925
(Fig. 88)

Description. Body outline shape oval to elongate, usually with broadly rounded posterior (but one species has narrowed posterior region), anterior always with conspicuous trichocyst wart projecting from the dorsal edge of the apical unciliated ridge. The anterior of the cell is truncated obliquely. Body ciliation uniform. Macronucleus either ovoid or elongate. Single posterior contractile vacuole.

Key to species. Kahl (1930–35).

Descriptions of species. Brodsky (1925), Kahl (1930).

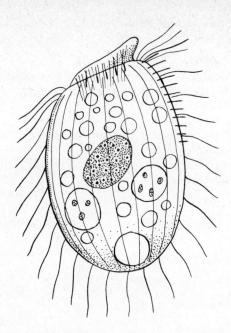

Fig. 87. *Spathidiodes* Kahl (after Kahl, 1930–35).

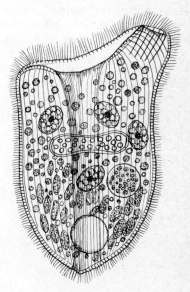

Fig. 88. *Spathidioides* Brodsky (after Kahl, 1930–35).

Spathidiosus Gajewskaja, 1933
(Fig. 89)

Description. Body shape bell-like, with broadly rounded posterior, open anterior end truncated transversely, rounded in cross-section. There is a very large open apical aperture occupying the entire anterior end of the body which is surrounded by a comparatively thick unciliated ridge. The body is metabolic with the anterior part around the oral aperture being rather more contractile than the posterior. Ciliation uniform in longitudinal kineties. There are groups of trichocysts scattered regularly throughout the cell just below the pellicle. Single spherical macronucleus. Single contractile vacuole. Planktonic in Lake Baikal feeding on diatoms.

Description of species. Gajewskaja (1933).

Spathidium Dujardin, 1841
(Fig. 90)

Spathidiosus Kahl, 1930

Spathidiopsis Kahl, 1926

Description. Body shape elongate, rounded in cross-section, posterior end bluntly pointed or rounded. Anterior region of body characteristically terminates obliquely but variable between transverse to longitudinal to the major body axis. There is always an unciliated apical ridge which is lined by trichocysts. The oral aperture is a slit lying along the length of the ridge. Ciliation uniform on both lateral surfaces in longitudinal parallel rows. Macronucleus highly variable, often elongate, ribbon-like or moniliform. Contractile vacuole single and terminal. Feeds on other ciliates.

Originally described by Dujardin (1841b).

Key to species. Kahl (1930–35).

Descriptions of species. Dragesco (1966b), Fryd-Versavel *et al.* (1976), Kahl (1930), Vuxanovici (1962a).

Morphological variability. Wenzel (1955), Woodruff and Spencer (1922).

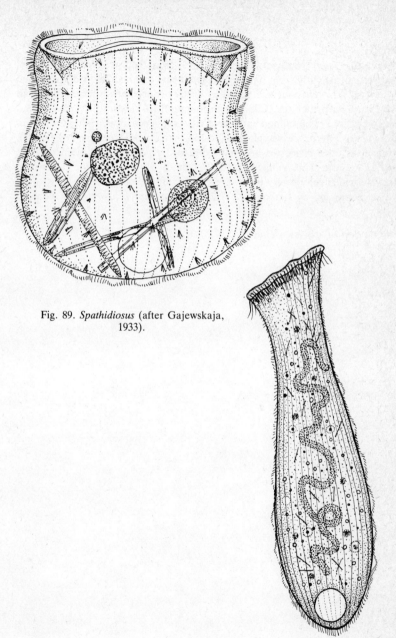

Fig. 89. *Spathidiosus* (after Gajewskaja, 1933).

Fig. 90. *Spathidium* (after Dragesco, 1966b).

Thysanomorpha Jankowski, 1967
(Fig. 91)

Description. Body outline shape oval with truncated anterior end. Apical region slants slightly from right to left and bears an unciliated ridge in which lies the slit-like oral aperture. Large cytopharynx occupies anterior quarter of cell, lined with trichites. Ciliation in longitudinal kineties. The dorsal and ventral edges are both lined with short processes each of which has the ability to extend. The processes have knob-like extremities and are armed with trichocysts. The processes are restricted to the posterior two-thirds of the body. Macronucleus elongate and C-shaped. Single posterior contractile vacuole which can occupy up to one-quarter of the body.

Genus erected by Jankowski (1967c) for *Legendrea bellerophon* Penard, 1914.

Description of species. Penard (1914).

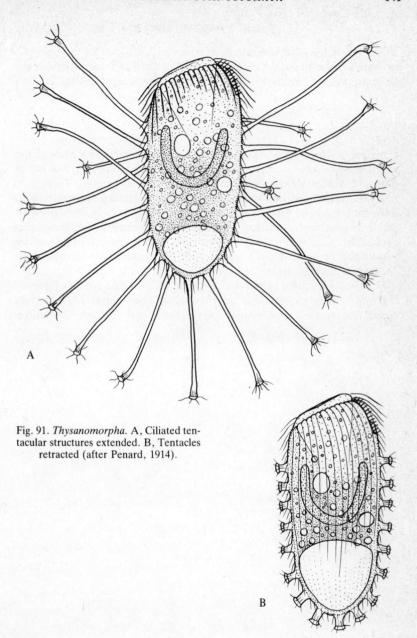

Fig. 91. *Thysanomorpha*. A, Ciliated tentacular structures extended. B, Tentacles retracted (after Penard, 1914).

Family TRACHELIIDAE

In this group the oral aperture is circular and is situated at the base of a proboscis. There are long prominent nematodesmata in the cytopharynx and there are many toxicysts present. Body is frequently large in size.

Branchioecetes Kahl, 1931
(Fig. 92)

Description. Body shape elongate, narrowing anteriorly to a blunt point, rounded posteriorly resembling shape of *Amphileptus* (p. 172) but not as flattened. Uniformly ciliated all over body in parallel kineties. There is an unciliated ridge lined with trichocysts leading to a rounded oral aperture (as in *Dileptus* (p. 144)) which is situated approximately halfway down the body on the convex ventral surface. Oral aperture supported by trichites. Many contractile vacuoles in rows along dorsal and ventral surfaces. Macronucleus moniliform or in scattered rounded parts. Ectocommensal on freshwater crustaceans such as *Asellus* and *Gammarus*.

Most easily confused with *Amphileptus* (p. 172) which is laterally flattened and has a long slit-like oral aperture along the ventral edge whereas *Branchioecetes* has a rounded oral aperture at the base of an unciliated ridge.

Key to species. Kahl (1930–35).

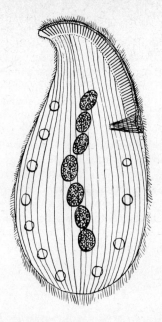

Fig. 92. *Branchioecetes* (composite from Kahl, 1930–35 and Penard, 1922).

Dileptus Dujardin, 1841
(Fig. 93)

Description. Body shape highly elongate, size ranges 100–1600 μm long, rounded in cross-section, with long highly mobile, contractile prehensile anterior neck region. Ventral surface of neck lined with trichocysts. Some species have a pointed tail region, others are rounded posteriorly. At the base of the neck lies the oral aperture which is supported by a cytopharyngeal basket of trichites (not always easily visible). The neck is often held extended, bent towards the dorsal surface. Trichocysts commonly present in cytoplasm and particularly down the unciliated ridge of the neck. Ciliation complete, in form of longitudinal kineties. Contractile vacuoles usually numerous in row along dorsal surface often with larger terminal vacuole. Contractile vacuoles rarely on both dorsal and ventral surfaces. Macronucleus highly variable from species to species but usually in two to many parts which may be distributed throughout the cytoplasm or may be moniliform, rarely single but when so is elongate. Originally described by Dujardin (1841a).

Myriokaryon Jankowski, 1973
(Fig. 94)

Description. Body shape elongate, worm-like, rounded in cross-section. Large (up to 1.3 mm long), sluggish and non-contractile. Apex bears two unciliated ridges between which lies a long slit. The parallel ridges extend a third of the length of the ventral surface where they join to surround the oral aperture. Trichocysts are abundant, concentrated beneath the ridges and extend past the oral region to the middle of the body. There is a single large terminal contractile vacuole and many supplementary vacuoles dispersed throughout the cell. Macronucleus fragmented (up to 2700 pieces) with numerous micronuclei. Planktonic, feeding on ciliates, algae and bacteria.

Jankowski (1973c) erected the genus for the species *Pseudoprorodon lieberkuhnii* (Bütschli, 1889) Kahl, 1930; *Spathidium gigas* Da Cunha, 1914 and *Cranotheridium elongatum* Penard, 1922.

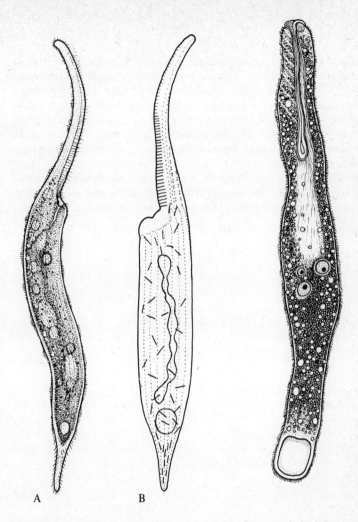

A   B

Figl 93. *Dileptus.* A, Detail drawn from life. B, Nuclear structure (after Grain and Golinska, 1970).

Fig. 94. *Myriokaryon* showing large numbers of macronuclei (after Jankowski, 1973c).

Paradileptus Wenrich, 1929
(Fig. 95)

Description. Body shape oval or conical, broad and obliquely truncated at the level of the oral aperture. There is a long, tapering, spiralling neck region at the anterior end of the body. In the centre of the neck lies a longitudinal unciliated ridge, armed with trichocysts, which winds anticlockwise with the neck down to the circular oral aperture which is supported by trichites. Ciliation is complete and uniform in longitudinal rows. The contractile vacuoles are numerous and lie beneath the pellicle all over the cell surface. Macronucleus moniliform.

May be distinguished from *Dileptus* (p. 144) by the spiralling neck, the way in which the unciliated ridge curves partially around the oral aperture (Fig. 36) and by the increased breadth of the body at the level of the oral aperture.

Descriptions of species. Canella (1951), Wenrich (1929b).

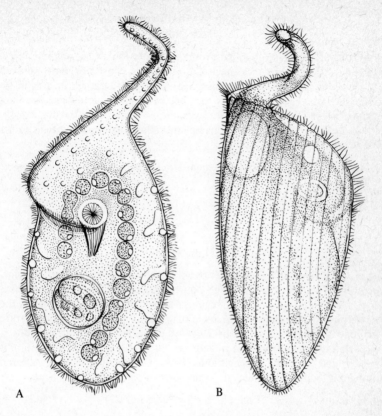

Fig. 95. *Paradileptus*. A, Ventral aspect (after Wenrich, 1929b). B, Dorsal aspect (from life).

Teuthophrys Chatton and Beauchamp, 1923
(Fig. 96)

Description. Shape irregular with rounded posterior part bearing three large anterior tapering arms or neck-like processes. Posterior part of body frequently contains green zoochlorellae symbionts. Oral aperture is in the shape of a triangular slit lying at the base of the three arms which are lined with trichocysts on their inner surfaces. Single terminal contractile vacuole. Macronucleus moniliform or elongate and ribbon-like. Single species genus. Feeds upon rotifers.

Morphology. Canella (1951), Clément-Iftode and Versavel (1968), Wenrich (1929a).

Trachelius Schrank, 1803
(Fig. 97)

Description. Body shape variable, ovoid to spherical bearing short anterior, finger-like process or snout which points towards the dorsal surface. Posterior rounded never with tail region. An unciliated ridge lies along the ventral face of the snout at the base of which lies a circular oral aperture surrounded by the ridge as in *Dileptus* (Fig. 36). The oral aperture is supported by trichites but the ridge is not lined with trichocysts. Ciliation is uniform and complete in longitudinal rows. Contractile vacuoles scattered throughout body. Macronucleus in one or two parts, usually elongate.

Descriptions of species. Kahl (1930–35), Penard (1922).

Morphology. Hamburger (1903).

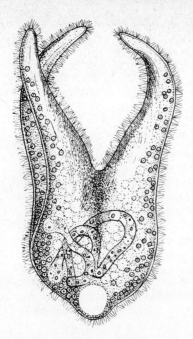

Fig. 96. *Teuthophrys* (after Wenrich, 1929a).

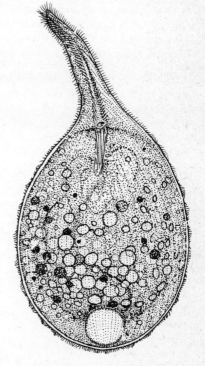

Fig. 97. *Trachelius* (drawn from life).

Family DIDINIIDAE

The oral aperture is apically located and the cytopharynx is reversible in some species. The somatic ciliature is reduced to one or more bands that encircle the typically barrel-shaped body.

Acropisthium Perty, 1852
(Fig. 98)

Description. Body shape ovoid to cylindrical, with pointed posterior region and short anterior snout-like process. Oral aperture apical on domed snout, cytopharynx armed with trichites. There is a ciliary band around the shoulder where the snout joins the body which is composed of about 20 short rows of cilia. The body behind the ciliary band is uniformly ciliated by about 20 longitudinal kineties which are usually well spaced, however the anterior half of three of the kineties are closely packed and these bear short clavate bristles representing the dorsal brush. Macronucleus spherical in posterior half of body. Contractile vacuole posterior. Single-species genus.

Description of species. Bohatier and Detcheva (1973).

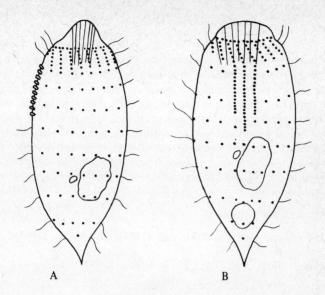

Fig. 98. *Acropisthium*. A, Lateral aspect. B, Dorsal aspect (after Bohatia and Detcheva, 1973).

Askenasia Blochmann, 1895
(Fig. 99)

Description. Body shape approximately pyriform, with larger, rounded posterior region and a smaller domed anterior region. Oral aperture apically situated, cytopharynx supported by trichites. There are two ciliary bands set close together in the anterior half of the cell. The anteriormost band is composed of short rows of short cilia projecting forward and the posterior-most band of longer rows of long cilia projecting backwards. In some species there is a circle of bristles (cirri) immediately behind the posterior ciliary band, these project stiffly out radially. In one species the posterior half of the body is sparsely ciliated. Macronucleus spherical to ovoid. Contractile vacuole terminal.

Key to species. Kahl (1930–35).

Descriptions of species. Fauré-Fremiet (1924), Gajewskaja (1933), Tamar (1973).

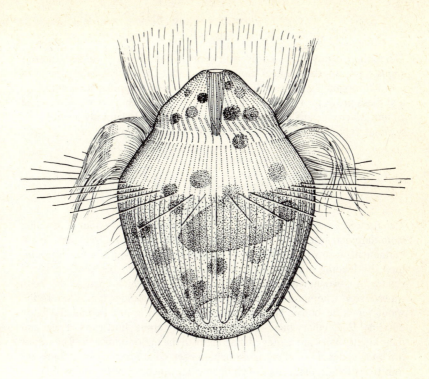

Fig. 99. *Askenasia* (after Fauré-Fremiet, 1924).

Choanostoma Wang, 1931
(Fig. 100)

Description. Body shape ovoid with short but pronounced domed anterior snout. Oral aperture apical mounted on snout. There is a single band of cirri-like organelles which encircles the cell where the snout joins the body and these project forward. Body uniformly ciliated behind cirri, with caudal cilium. Macronucleus moniliform. Contractile vacuole laterally located.

Single report by Wang (1931) in Shanghai area, China.

Didinium Stein, 1859
(Fig. 101)

Description. Body barrel-shaped with short cone-shaped snout protruding from flattened anterior region, posterior broadly rounded. Oral aperture not permanent, forms only when ingesting prey which is typically *Paramecium*; then the feeding oral aperture becomes highly expandable. Body ciliation reduced to two narrow bands of closely-set cilia which encircle the body transversely. The anterior band is located at the shoulder region where the snout joins the body. The posterior ciliary band is just behind the body mid-line. Each ciliary band is composed of numerous short rows of cilia and their kinetosomes are arranged diagonally to the major body axis. The rest of the body is devoid of cilia except behind each ciliary band, where there are four to six longitudinal rows of clubbed cilia, which are not easily visible but may be shown by silver impregnation methods or electron microscopy. Macronucleus sausage to horse-shoe shape. Contractile vacuole posterior.

Key to species. Kahl (1930–35) – who included *Monodinium* (p. 160) and *Dinophrya* (p. 156) in the genus.

Descriptions of species. Stein (1859a), Wessenberg and Antipa (1969).

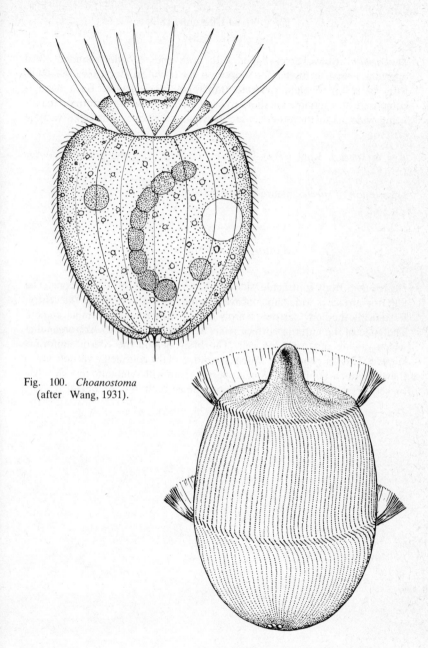

Fig. 100. *Choanostoma*
(after Wang, 1931).

Fig. 101. *Didinium* (after Wessenberg
and Antipa, 1969).

Dinophrya Bütschli, 1889
(Fig. 102)

Description. Body barrel-shaped with both ends broadly rounded. Oral aperture apical, cytopharynx supported by trichites. Body ciliation takes form of seven to eight separate transverse ciliary bands. Each band is composed of a number of short rows of cilia, the two anteriormost bands being wider than the others. Macronucleus spherical. Contractile vacuole posterior.

Key to species. Kahl (1930–35) – who includes it in the genus *Didinium* (p. 154).

Description of species. Fauré-Fremiet (1924).

Liliimorpha Gajewskaja, 1928
(Fig. 103)

Description. Body contractile, shaped like an upturned short, wide cone. The anterior surface is wide and concave bearing the oral aperture in the centre. Posteriorly the body narrows sharply to a broadly rounded terminal region. The edge of the anterior surface bears a single band of cirri-like organelles which radiate out from the cell. The body behind the cirri is uniformly ciliated. The macronucleus, which is ovoid, and the contractile vacuole are in the anterior part of the cell. Single species genus with symbiotic zoochlorellae present. Has been reported on two occasions, both in Lake Baikal.

Descriptions of species. Gajewskaja (1928, 1933).

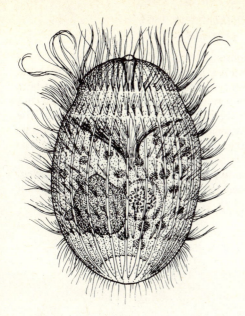

Fig. 102. *Dinophrya* (after Fauré-
Fremiet, 1924).

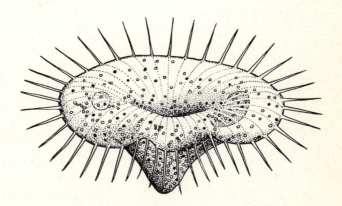

Fig. 103. *Liliimorpha* (after Gajewskaja,
1933).

Mesodinium Stein, 1863
(Fig. 104)

Description. Body shape pyriform but divided into two sections by a waist-like furrow. Anterior narrows apically, posterior broadly rounded. Oral aperture apical, cytopharynx with eight to twelve protruding trichites which are forked at the anteriormost end. Somatic ciliation reduced to two bands of cilia which arise from the central body furrow. Both bands arise from short oblique rows of kinetosomes, the posterior band consists of rows of paired kinetosomes, the anterior rows have single kinetosomes. The components of the anterior ciliary band are stiff and bristle-like, those of the posterior band are membranelle-like and project backwards. Macronucleus centrally located. Contractile vacuoles have been reported in both posterior and mid-lateral positions.

Key to species. Kahl (1930–35).

Descriptions of species. Borror (1972), Tamar (1971).

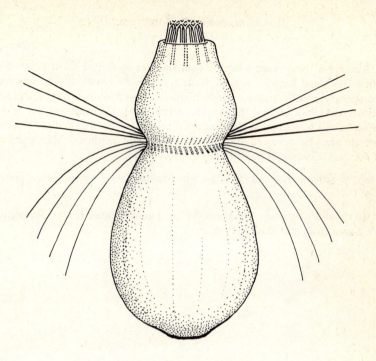

Fig. 104. *Mesodinium* (after Borror; 1972).

Monodinium Fabre-Domergue, 1888
(Fig. 105)

Description. Body barrel-shaped with short cone-shaped snout protruding from flattened anterior region. Posterior broadly rounded. Oral aperture temporary, apically situated. Body ciliation reduced to single narrow band of closely-set cilia encircling the shoulder region where the snout joins the body. Rest of cell free from cilia except for four to six short longitudinal kineties which may be revealed by silver impregnation methods. Macronucleus C-shaped. Contractile vacuole posterior.

Key to species. Kahl (1930–35) – who includes it in the genus *Didinium* (p. 154).

Descriptions of species. Dragesco (1970), Fauré-Fremiet (1945), Rodrigues de Santa Rosa and Didier (1976).

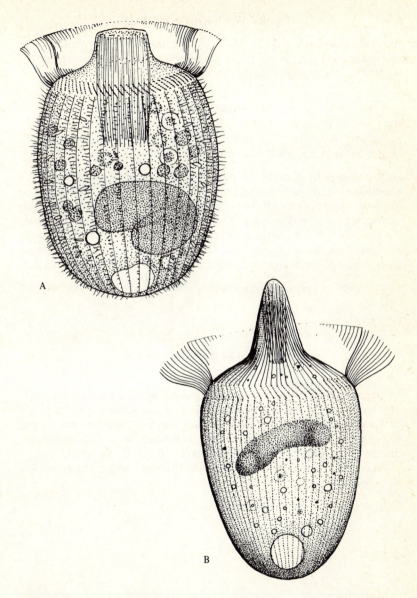

Fig. 105. *Monodinium*. A, after Fauré-Fremiet (1924). B, After Dragesco (1970).

Family ACTINOBOLINIDAE

The oral aperture is apical and the somatic ciliature is uniform. The distinguishing feature of the family is the presence of non-suctorial tentacles which have toxicysts associated with them.

Actinobolina Strand, 1928
(Fig. 106)

Actinobolus Stein, 1867

Description. Body shape ovoid, posterior always broadly rounded, anterior sometimes slightly narrowed. Oral aperture apically located, supported by cytopharyngeal basket of trichites. Body covered in uniform ciliation, either in longitudinal or oblique rows. There are many extensible tentacles among the cilia which may be retracted or extended to about twice the body width. Macronucleus usually elongate and ribbon-like but may be in two spherical parts. Single terminal contractile vacuole. Carnivorous, feeds on rotifers.

Descriptions of species. Fauré-Fremiet (1924), Kahl (1930–35), Wenrich (1929c).

Belonophrya André, 1914
(Fig. 107)

Description. Body shape ovoid, bearing slight apical protruberance which denotes the site of the oral aperture. There is a prominent cytopharyngeal basket of trichites. Body covered in short cilia and there are many rigid, non-expansible tentacles scattered over its surface. Macronucleus sausage-shaped. Single posterior contractile vacuole.

Most easily confused with *Actinobolina* (above) which has expansible tentacles.

Description of species. André (1914).

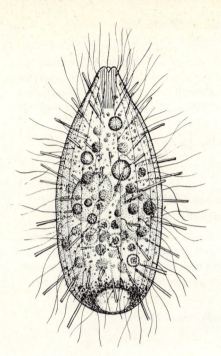

Fig. 106. *Actinobolina* (after Fauré-
    Fremiet, 1924).

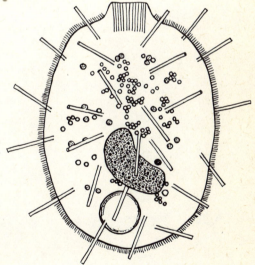

Fig. 107. *Belonophrya* (after André,
    1914).

Dactylochlamys Lauterborn, 1901
(Fig. 108)

Description. Body shape irregularly elongated ovoid with posterior end drawn out into a tail region. Covered in eight to twelve crenated ridges which spiral down around the body. Long cilia and retractile capitate tentacles are alternately located along the ridges. Macronucleus spherical and centrally located. Single contractile vacuole.

Descriptions of species. Kahl (1930–35), Lauterborn (1901).

Enchelyomorpha Kahl, 1930
(Fig. 109)

Description. Body shape elongate, conical, narrowing anteriorly, posterior rounded, dorso-ventrally flattened. Body deeply furrowed transversely from which the cilia emerge. Anterior half of body has short non-retractile capitate tentacles scattered over it. Macronucleus spherical, centrally located. Single large terminal contractile vacuole.

Descriptions of species. Kahl (1930–35), Smith (1899).

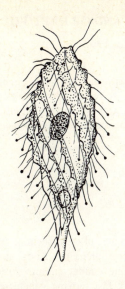 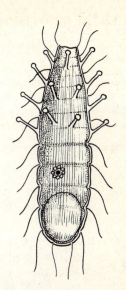

Fig. 108. *Dactylochlamys* (after Kahl, 1930–35).   Fig. 109. *Enchelyomorpha* (after Kahl, 1930–35).

INCERTAE SEDIS PROSTOMATIDA AND/OR HAPTORIDA

Baznosanuia Tucolesco, 1962
(Fig. 110)

Description. Body outline shape rectangular with parallel sides, circular in cross-section. Entire body covered by double spiral striations which cross each other obliquely forming diamond-like pattern on surface. Contractile vacuole centrally located. Found in a cave pool, single record (Romania).

Description of species. Tucolesco (1962).

Celeritia Tucolesco, 1962
(Fig. 111)

Description. Body spherical to ovoid, rigid, circular in cross-section. Apical rounded oral aperture leading to conical cytopharynx. Body striated obliquely, few long cilia on body, single long caudal cilium. Posterior contractile vacuole. Found in cave pool, very fast moving, single record (Romania).

Description of species. Tucolesco (1962).

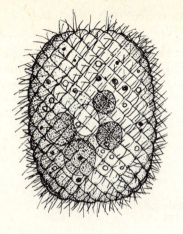

Fig. 110. *Baznosanuia* (after Tucolesco, 1962).

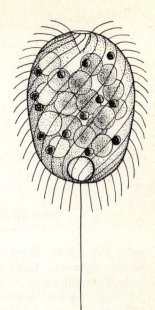

Fig. 111. *Celeritia* (after Tucolesco, 1962).

Pseudoenchelys Tucolesco, 1962
(Fig. 112)

Description. Body shape elongate, metabolic, rounded posterior, anterior truncated transversely. Covered in longitudinal furrows. Oral aperture rounded, surrounded by forward pointing spines and armed internally with trichocysts. Uniformly ciliated. Macronucleus C-shaped. Posterior contractile vacuole located on one side. Found in cave pool feeding on ciliates, single record (Romania).

Description of species. Tucolesco (1962).

Racovitzaia Tucolesco, 1962
(Fig. 113)

Description. Body outline shape irregularly oval, rounded ends with anterior projection on right side. There is a deep furrow beneath this projection. Ciliation uniform with a long caudal cilium. Macronucleus centrally located, spherical. Single terminal contractile vacuole. Found in cave pool, single record (Romania).

Description of species. Tucolesco (1962).

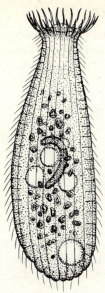

Fig.   112.   *Pseudoenchelys*
(after Tucolesco, 1962).

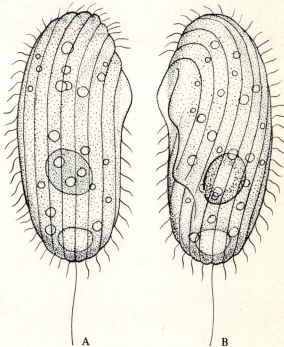

A   B

Fig. 113. *Racovitzaia* A, Dorsal view B,
Ventral surface (after Tucolesco, 1962).

Order PLEUROSTOMATIDA

Members of this order possess a slit-like oral aperture that is located along the ventral edge of the laterally compressed body. They are often large and voracious carnivores.

Family AMPHILEPTIDAE

With the characters of the above order.

Acineria Dujardin, 1841
(Fig. 114)

Description. Body shape elongate, laterally flattened with rounded posterior, anterior narrows to blunt point. Oral aperture is a slit lying along the obliquely situated apical edge. There is a line of trichocysts below the oral edge. Many cilia in rows on right (lower) surface which continue past the dorsal edge onto the dorsal side of the left (upper) surface. The ventral half of the left surface is unciliated. Single terminal contractile vacuole. Macronucleus in two ovoid parts with micronucleus between them. Single-species genus first described by Dujardin (1841b). Feeds upon small ciliates such as *Colpidium* and *Cyclidium*.

Descriptions of species. Kahl (1930–35), Bick (1972).

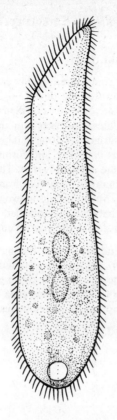

Fig. 114. *Acineria* (after Kahl, 1930–35).

Amphileptus Ehrenberg, 1830
(Fig. 115)

Hemiophrys Wrzesniowski, 1870

Description. Body laterally compressed, highly elongate with anterior neck-like region which bends towards the dorsal edge. Oral aperture a slit on convex edge of neck region, extends less than halfway down the body. Ciliation present on both lateral surfaces although there is a tendency to some reduction on the left surface resulting in it being difficult to distinguish. Ciliation on right surface is extensive and forms longitudinal rows which converge on each other in the anterior region. There is a distinctive area of cilia along the oral slit forming a mane-like brush. Trichocysts commonly present particularly in neck. Macronucleus in two to four spherical parts with single micronucleus placed between macronuclei. Many contractile vacuoles present, usually lying along dorsal and ventral edges.

Taxonomy. The recommendations and findings of Canella (1960) and Fryd-Versavel *et al.* (1976) have been adopted here so that the genus *Hemiophrys* has been submerged into the above genus. This is because cilia do appear to be present on the left surface even though difficult to resolve without the use of silver impregnation techniques.

Key to species. Kahl (1930–35).

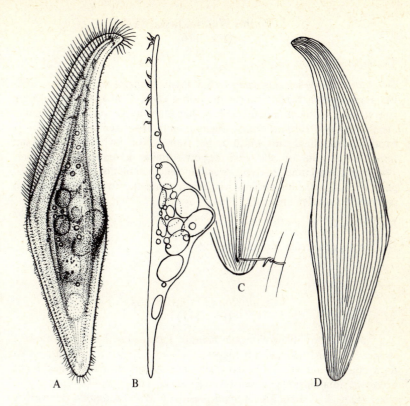

Fig. 115. *Amphileptus*. A, Whole animal. B, Lateral view, animal distended by prey. C, Method of attachment to peritrich stalk (all drawn from life). D, Kineties on upper surface (after Canella, 1960).

Litonotus Wrzesniowski, 1870
(Fig. 116)

Lionotus Bütschli, 1887

Description. Body laterally compressed, highly elongate with anterior neck-like region which bends towards the dorsal edge. Oral aperture a slit, on convex edge of neck extending less than halfway down body. Ciliation present on both lateral surfaces. Ciliation on right surface takes the form of parallel longitudinal rows which do not converge on each other. There are some longer cilia on the neck region forming a mane-like structure. Trichocysts sometimes present. Macronucleus commonly in two spherical parts with single micronucleus wedged between the two. One to several contractile vacuoles present.

Key to species. Kahl (1930–35).

Description of species. Vuxanovici (1960).

Loxophyllum Dujardin, 1841
(Fig. 117)

Description. Body shape elongate, laterally flattened with flat right (lower) surface and partly domed left (upper) surface. Anterior region narrows to form a blunt point which is held dorsally. The oral aperture takes the form of a long slit situated along the ventral edge. There is no mane-like brush of cilia along the neck as there is in *Amphileptus* (p. 172) and *Litonotus* (above). A flattened non-granular band which is often striated transversely by many trichocysts runs down the entire length of the ventral edge. The dorsal edge either resembles the ventral edge with a band of trichocysts or more commonly there is a series of trichocyst warts at intervals along its length. There are many ciliary rows on the right (lower) surface and fewer on the left (upper). There is usually a large terminal contractile vacuole either with or without smaller vacuoles along the dorsal and/or ventral edges, rarely with long dorsal serving canal. Macronucleus ovoid in one, two or many moniliform pieces. Carnivorous, often feeding upon rotifers and other ciliates.

First described by Dujardin (1841b).

Key to species. Kahl (1930–35).

Descriptions of species. Vuxanovici (1959a).

Morphology and ultrastructure. Fryd-Versavel *et al.* (1976), de Puytorac and Rodrigues de Santa Rosa (1976).

Fig. 116. *Litonotus* (composite after Roux, 1901; Canella 1960 and Bick, 1972).

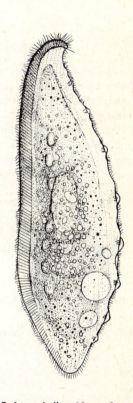

Fig. 117. *Loxophyllum* (drawn from life).

Keys and descriptions of genera of Vestibulifera

Subclass VESTIBULIFERA

The vestibuliferans characteristically have a cytostome at the base of a vestibular cavity which is an oral depression lined with more or less distinctive cilia that are derived from somatic cilia. The cytopharyngeal apparatus is reminiscent of the rhabdos type and toxicysts are rarely present. Until recently there were three orders in the subclass but the work of Fernandez-Galiano (1978) has resulted in the erection of a new order, the Bursariomorphida. Thus there are now three orders with free-living freshwater representatives and a fourth which is completely parasitic.

Key to genera of loricate Vestibulifera

1. Branched or unbranched lorica. Contractile vacuole situated in extreme anterior region of body **2**

   Lorica always unbranched. Contractile vacuole in posterior half of body .... *Cyrtolophosis* (p. 198)

2. Body heavily ciliated, often with anterior collar-like protruberance .... *Maryna* (p. 182)

   Body with only four rows of cilia, never with anterior collar .... *Mycterothrix* (p. 184)

Key to genera of free-swimming Vestibulifera

1. Contractile vacuole in apical region of body **2**

  Contractile vacuole in posterior or central region of body, when numerous they are arranged around the cell perimeter **3**

2. Cell with prominent anterior collar-like peristome, ends of body rounded .... *Maryna* (p. 182)

  Cell without collar, ends of body pointed .... *Mycterothrix* (p. 184)

3. Body covered with many ciliary rows, when caudal cilia present they are numerous **4**

  Body with only four ciliary rows, single caudal cilium present (Fig. 31A) .... *Trimyema* (p. 188)

4. A pellicular rib bearing long, closely set cilia, spirals from oral region towards posterior. Group of caudal cilia present **5**

  Without a pellicular rib bearing cilia, caudal cilia never present but in one genus there are caudal cirri **6**

5. Pellicular rib spirals from an almost apical oral aperture and terminates in a transverse ring (Fig. 31C). Many trichocysts present .... *Trichospira* (p. 186)

  Pellicular rib spirals from laterally placed oral aperture and does not terminate in a transverse ring (Fig. 31B). Trichocysts absent .... *Spirozona* (p. 186)

6. Outline body shape reniform or with rostrum **7**

  Outline body shape oval or pyriform and may be truncated anteriorly. Never with a rostrum **19**

7. Body bends towards or has a notch in the right edge **8**

  Body bends towards or has a notch in the left edge **9**

8. Deep notch on right edge from which arises a deep furrow in the shape of a walking stick on the dorsal surface .... *Plagiopyla* (p. 184)

  Body with short anterior rostrum which bends to the right. Dorsal surface without furrow but with longitudinal ribs .... *Rigchostoma* (p. 216)

9. Terminal contractile vacuole with long serving canals **10**

  Contractile vacuole without canals but some genera have vesicles surrounding them **11**

10. Body twisted with an obliquely winding groove. Contractile vacuole with several radiating serving canals .... *Tillina* (p. 210)

    Body not twisted, without groove. Contractile vacuole with two serving canals .... *Bryophrya* (p. 192)

11. Body regularly curved to left, without marked indentation or notch on left edge **14**

    Body with marked indentation or notch on left edge **12**

12. With greatly enlarged vestibulum which occupies two-thirds of body. Contractile vacuole posterior but not terminal ....
    .... *Bresslaua* (macrostome) (p. 190)

    Without greatly enlarged vestibulum, contractile vacuole terminal **13**

13. Viewed orally the apex of the triangular-shaped oral aperture is directed posteriorly (Fig. 118A) .... *Bresslaua* (microstome) (p. 190)

    Viewed orally the apex of the triangular-shaped oral aperture is directed anteriorly (Fig. 118B) .... *Colpoda* (p. 196)

14. Posterior region with two to twenty-five tube-like structures (Fig. 16F) .... *Cirrophrya* (p. 194)

    Posterior without tube-like structures **15**

15. Contractile vacuole terminal. Macronucleus spherical **16**

    Contractile vacuole in central region. Macronucleus elongate
    .... *Bursostoma* (p. 194)

Silver Preparation necessary for next three stages. Several methods have been successfully applied

16. On the left of the oral aperture there is a line of membranelles each composed of a double row of kinetosomes which projects forward well beyond the oral area (Fig. 119A,B) **17**

    On the left of the oral aperture there is a line of membranelles each composed of a double row of kinetosomes which either does not or only just projects beyond the oral area (Fig. 119C,D) **18**

17. Line of membranelles begins at posterior end of oral area and extends forward (Fig. 119B) .... *Rostrophyra* (p. 208)

    Line of membranelles begins at anterior end of oral area and extends forward (Fig. 119A) .... *Kuklikophrya* (p. 202)

18. Line of membranelles may just extent forward beyond oral area. Body evenly ciliated on both sides (Fig. 119C) ...... *Woodruffia* (p. 212)

    Line of membranelles does not extend beyond oral area. Body with reduced ciliation on left side (Fig. 119D) ..... *Platyophrya* (p. 204)

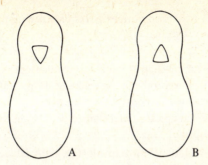

Fig. 118. Shape of oral aperture in two colpodid ciliates. A, *Bresslaua* microstome. B, *Colpoda* (after Stout, 1960).

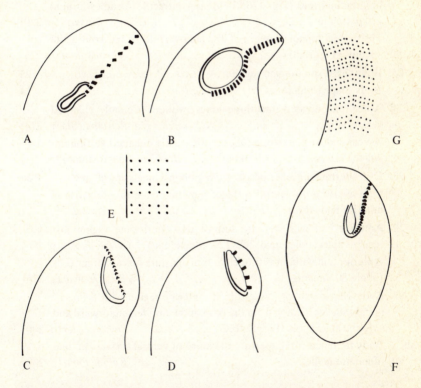

Fig. 119. Oral structures in some vestibuliferan ciliates. A, *Kuklikophrya* (after Njiné, 1979a). B, *Rostrophrya* (after Njiné, 1979b). C, *Woodruffia* (after Grolière, 1975). D, *Platyophrya* (after Foissner, 1978a). E, Membranelles in above mentioned genera arise from paired ciliary rows. F, *Puytoraciella* (after Njine, 1979b). G, Membranelles of latter genus arise from triple ciliary lines.

19. Rare form with four posterior cirri. Has only been reported from
Lake Baikal .... *Sulcigera* (p. 218)
Without cirri **20**

20. Peristome is a funnel-shaped depression which opens anteriorly
and except for a ventral slit is enclosed (Fig. 120). Always
transversely truncated anteriorly **21**
Peristome never an apical funnel-shaped depression. Body never
truncated anteriorly **23**

21. Base of peristomial funnel turns to left (Fig. 120A,B) **22**
Base of peristomial funnel turns to right (Fig. 120C)
.... *Bursaridium* (p. 222)

22. Peristomial depression divided internally into two parts by a
longitudinal fold (Fig. 120A). Many contractile vacuoles around
cell perimeter .... *Bursaria* (p. 220)
Peristomial funnel simple, not divided by a fold. Single contractile
vacuole (Fig. 120B) .... *Thylakidium* (p. 223)

23. Body shape smooth and regular **25**
Body shape irregular ....,.................. **24**

24. Apical end divided into three lobes, without a greatly enlarged
vestibulum .... *Opisthostomatella* (p. 214)
Apical end without three lobes, with greatly enlarged vestibulum
which occupies two-thirds of body ..... *Bresslaua* (macrostome) (p. 190)

25. Oral aperture apical or peristome region just meets apical pole ... **26**
Oral aperture on ventral surface, sometimes central otherwise in
anterior body third **27**

26. Anterior end with definite tuft of cilia, peristome region just
reaches apical pole, small cells (20–40 μm long) ... *Cyrtolophosis* (p. 198)
Anterior end without tuft of cilia, oral aperture apical, larger cells
(about 80 μm long) .... *Orcavia* (p. 216)

27. Body ellipsoidal, oral aperture in anterior region with a line of
membranelles leading from the base of the oral region forward and
over the apical pole (Fig. 119F,G) .... *Puytoraciella* (p. 206)
Body flattened, oral aperture in centre of ventral surface, no line
of membranelles .... *Kalometopia* (p. 200)

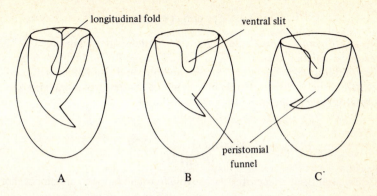

Fig. 120. Distinguishing features of three vestibuliferan ciliate genera. A, *Bursaria*. B, *Thylakidium*. C, *Bursaridium*.

Sessile Vestibulifera

The single genus *Grandoria* is the only sessile representative of the Vestibulifera. It attaches itself to substrata in soil by a bunch of posterior gelatinous fibres. This is a rare genus (p. 214).

Order TRICHOSTOMATIDA

These are simple forms resembling the prostomatine gymnostomes. The vestibular ciliation is simple and is derived during fission from the terminal parts of the somatic kineties. The oral aperture and vestibulum are apically located in the free-living forms but sometimes posteriorly in the endozooic forms. Several genera produce gelatinous loricas.

Maryna Gruber, 1879
(Fig. 121)

Description. Body shape irregularly ovoid to round but with prominent short collar-like peristome projecting from anterior end of cell. The vestibulum takes the form of a longitudinally arranged slit on the ventral surface. Body ciliation complete, those at the anterior sometimes being longer than those on the rest of the body. Cell may be found in a lorica or swimming free. When in the gelatinous lorica the cell usually partially projects from it. Lorica may be short but often long and always tubular, either solitary or colonial in branched lorica. Contractile vacuole always apical, sometimes with two short serving canals. Macronucleus rounded and centrally positioned. With one exception, those species originally placed in the genus *Mycterothrix* (p. 184) have been transferred to *Maryna* (see Dingfelder, 1962; Jankowski, 1964).

Descriptions of species. Buitkamp (1975), Dingfelder (1962), Gelei (1950).

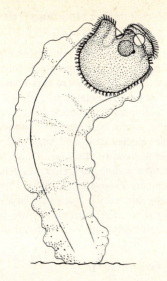

Fig. 121. *Maryna* (composite after Gelei, 1950).

Mycterothrix Lauterborn, 1898
(Fig. 122)

Description. Body ovoid with pointed ends, circular in cross-section. Body ciliation reduced to four concentric rings which encircle the body, and in addition there appear to be several rows of cilia above the vestibulum as in *Trimyema* (p. 188). May inhabit gelatinous lorica. Contractile vacuole in apical region of body. Macronucleus rounded, located anteriorly.

The diagnosis of the genus has changed considerably since that of Lauterborn (1898). The species described since that time have now been transferred to the genus *Maryna* (p. 182) (see Dingfelder, 1962; Jankowski, 1964) and only *Mycterothrix acuminata* Gellért, 1955 has been retained even though it strongly resembles *Trimyema* (p. 188). However, in view of the aberrant position of the contractile vacuole (Jankowski, 1964) and the presence of four rather than three ciliary rings, we have not followed the suggestion of Jankowski (1964) and have not transferred it to the genus *Trimyema*.

Description of species. Gellért (1955).

Taxonomy. Dingfelder (1962), Gelei (1950), Jankowski (1964).

Plagiopyla Stein, 1860
(Fig. 123)

Description. Body laterally compressed, oval in outline with distinct notch in right hand edge (ventral) in anterior body quarter. Distinct vestibular area in form of a transversely arranged deep ciliated furrow which meets the notch on the right (ventral) edge. The furrow travels approximately halfway across the body on the apparent ventral (left) surface. On the apparent dorsal (right) surface there is a channel-like transversely striated structure which extends initially up from the vestibulum for a short distance before sharply bending down to travel posteriorly along the right edge. There are many ciliary meridians which originate from the vestibular furrow. Body entirely ciliated. Large terminal contractile vacuole. Macronucleus spherical or ovoid, centrally placed. Sapropelic.

Key to species. Kahl (1930–35).

Descriptions of species. Jankowski (1964), Stein (1860b).

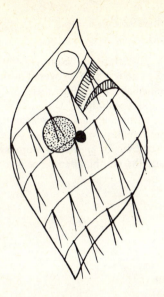

Fig. 122. *Mycterothrix* (after Gellért, 1955).

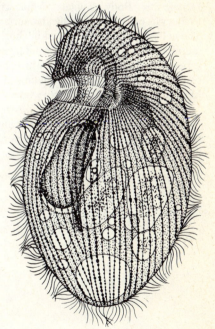

Fig. 123. *Plagiopyla* (composite after Jankowski, 1964 and Wetzel, 1928).

Spirozona Kahl, 1926
(Fig. 124)

Description. Body elongate ovoid with posterior end bluntly pointed. Opening to vestibulum in anterior quarter of body which is covered in short cilia. There is a tuft of caudal cilia and a pellicular rib, bearing a series of long closely-set cilia, which originates near the vestibulum. The rib spirals clockwise down the body but does not terminate in a transverse ring. Macronucleus ovoid, contractile vacuole terminal.

Most easily confused with *Trichospira* (below). In *Spirozona* there is no transverse pellicular ring, no trichocysts and the vestibular entrance is more laterally positioned than in *Trichospira*.

Description of species. Kahl (1926).

Taxonomy. Jankowski (1964).

Trichospira Roux, 1899
(Fig. 125)

Description. Body elongate ovoid, circular in cross-section, with many trichocysts below entire pellicular surface. Opening to vestibulum subapical. The cell is completely covered in short cilia in densely-packed longitudinal meridians, with a tuft of long caudal cilia. There is a pellicular rib bearing a series of long closely-set cilia which begin near the vestibulum and spiral clockwise around the body ending in a transverse ring in the posterior. Macronucleus ovoid, contractile vacuole in posterior.

Most easily confused with *Spirozona* (above). *Trichospira* possesses trichocysts, the entrance to the vestibulum is very close to the body apex and the pellicular rib ends in a transverse ring. *Spirozona* does not have these features.

Descriptions of species. Jankowski (1964), Kahl (1930–35).

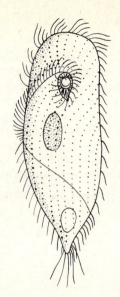

Fig. 124. *Spirozona* (after Kahl, 1926).

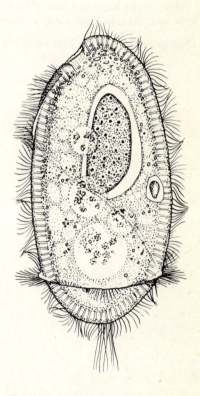

Fig. 125. *Trichospira* (after Jankowski, 1964).

Trimyema Lackey, 1925
(Fig. 126)

Description. Body outline shape irregular ovoid to reniform, circular in cross-section. Somatic ciliation reduced to three concentric rings encircling the body in the anterior region and a single caudal cilium. The anteriormost ciliary ring, near the vestibulum, branches into four ciliary stripes transversely above the vestibulum thus resembling the perizonal stripe in *Metopus* (see Part II). Contractile vacuole in anterior half of body but never apical. Macronucleus large in anterior half of body.

Most easily confused with *Mycterothrix* (p. 84) which has an apical contractile vacuole, four somatic ciliary rows and is without a caudal cilium.

Descriptions of species. Jankowski (1964), Klein (1930), Lackey (1925).

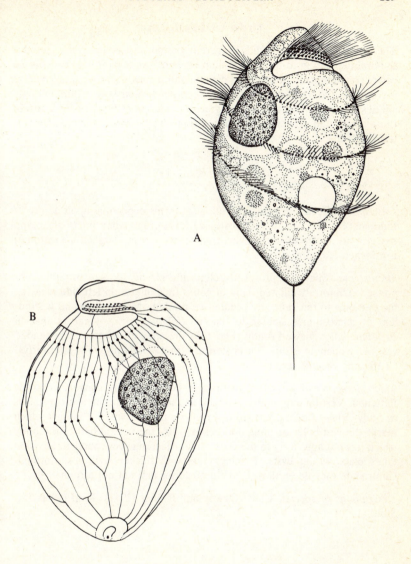

Fig. 126. *Trimyema*. A, Whole animal. B, Silver impregnated specimen (after Jankowski, 1964).

Order COLPODIDA

The oral ciliation of colpodids is currently under discussion, some consider the oral depression to be a vestibulum while others suggest it to be a buccal cavity containing membranelles ('cirromembranelles'). The somatic ciliation is complete and frequently the cilia arise in pairs. A striking feature of many species in the group is their high degree of asymmetry whose torsion is lost in those which undergo division within a reproductive cyst. Stomatogenesis is telokinetal.*

Bresslaua Kahl, 1931
(Fig. 127)

Description. Occurs in two morphological forms depending upon nature of food supply. The smaller (microstome) form grows when micro-algae and bacteria are the food source; this transforms into the larger (macrostome) form when fed upon ciliates.

Microstome. Shape reniform to ellipsoid, plastic, dorso-ventrally compressed. Uniformly ciliated, cilia in pairs. Indentation on left side marking the entrance to the ventrally situated vestibulum. Morphology and vestibulum of microstomes similar to that of *Colpoda* (p. 196) except that in *Bresslaua* when viewed orally (Fig. 118) the apex of the triangular-shaped vestibular opening is directed posteriorly (in *Colpoda* in points anteriorly). Contractile vacuole terminal.

Macrostome. Oval in outline with rigid body, dorsally convex, ventrally flattened. Vestibulum greatly enlarged so that it occupies almost two-thirds of body. Viewed dorsally there is a cleft on the left anterior margin which marks the vestibular opening and below it lies the ventral half of the body which juts out anteriorly to the dorsal half. Macronucleus rounded. Division takes place within cysts. Resting cysts with thicker walls also formed. Contractile vacuole on dorsal side in posterior half but not terminal.

Descriptions of species. Claff, Dewey and Kidder (1941), Kahl (1930–35), Stout (1960).

---

* Since this *Synopsis* went to press an important paper concerning colpodid ciliates has been published by Foissner, W. (1980). [Colpodide Ciliaten (Protozoa: Ciliophora) aus alpinen Böden. *Zool. Jb.* (Syst) **107**(3), 391–432]. The following new genera are included, *Grossglockneria* Foissner, 1980; *Nivaliella* Foissner, 1980; *Pseudocyrtolophosis* Foissner, 1980 and *Pseudoplatyophrya* Foissner, 1980. Descriptions of species of the following genera also included: *Colpoda*, *Cyrtolophosis*, *Paracolpoda* [see Lynn, D.N. (1978). *Diss. Abstr. Int.* **38** (9)], *Platyophrya* and *Woodruffia*.

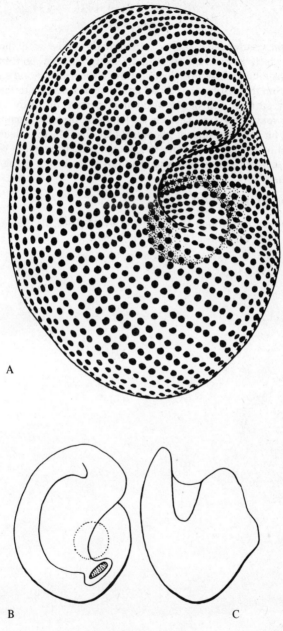

Fig. 127. *Bresslaua.* A, Ventral aspect of a silver impregnated microstome specimen. B, C, Ventral and lateral aspects of macrostome specimens (after Stout, 1960).

Bryophrya Kahl, 1931
(Fig. 128)

Description. Body shape oval to reniform in outline with short anterior rostrum which is directed to its left. Body ciliation uniform in many longitudinal/oblique rows originating from a distinct pre-oral suture which extends from the vestibular aperture on the ventral surface to the midline of the dorsal surface. Lip of the vestibular aperture distinctly U-shaped in centre of ventral surface. There is a posteriorly directed membranelle-like structure on the left side of the vestibular lip. Macronucleus spherical, equatorial to animal's right with two micronuclei. Contractile vacuole with two long serving canals which extend a third of the length of the body along its margins.

Key to species. Kahl (1930–35).

Taxonomy. Grain, Iftode and Fryd-Versavel (1980), de Puytorac, Perez-Paniagua and Perez-Silva (1979), Stout (1960).

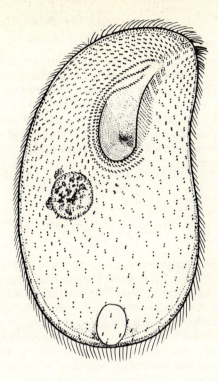

Fig. 128. *Bryophrya,* ventral aspect (after Puytorac *et al.*, 1979).

Bursostoma Vörösváry, 1950
(Fig. 129)

Description. Body shape ellipsoid to reniform, uniformly ciliated with many longitudinal/oblique rows which originate around the edge of the vestibular opening. The vestibulum is spherical and well developed opening via a figure-of-eight slit. The vestibulum is ciliated internally and contains two large membranelles. The cytostome lies at the bottom of the vestibulum. Macronucleus large and sausage-shaped, single small rounded micronucleus. Single contractile vacuole surrounded by several small serving canals.

Description of species. Vörösváry (1950).

Cirrophrya Gellért, 1950
(Fig. 130)

Description. Shape elongate reniform to pyriform, highly metabolic (see *Glossary*) with very short left anterior rostrum bearing an apical oval oral aperture. On the right of the aperture there is a C-shaped double row of cilia (paroral cilia) and there are about twelve groups of cilia on the left (adoral cilia). Body cilia uniform, arranged in longitudinal/oblique rows of cilia. At the posterior end of the body there are two to twenty-five curious tube-like structures. Contractile vacuole posterior. Macronucleus rounded, centrally placed, micronucleus within outer macronuclear membrane. Division takes place in cysts. Most easily confused with *Platyophrya* (p. 204) which does not have the posterior tubes.

Description of species. Gellért (1950a).

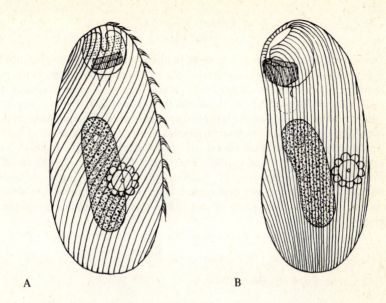

A   B

Fig. 129. *Bursostoma* (after Vörösváry, 1950). A, Lateral view. B, Ventral view.

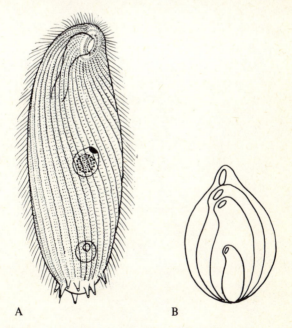

A   B

Fig. 130. *Cirrophrya* A, Whole animal. B, Diagram to show variation in size and shape
(after Gellért, 1950a).

Colpoda Müller, 1773
(Fig. 131)

Description. Body distinctly reniform in shape, dorso-ventrally flattened. Right body edge strongly convex, left body edge concave often appearing as though a bite had been taken from it. A shallow diagonal somatic groove (not easily visible) originating on the dorsal surface travels round left side to entrance of vestibulum on the flattened ventral surface. Ciliation uniform in longitudinal or oblique orientated grooves. Several notches which denote ciliary grooves often visible on preoral part of left body edge. Caudal cilia may be present in some species. There is a horse-shoe shaped arc of closely-set cilia on the right of the vestibular entrance. Single rounded macronucleus with one, two or three micronuclei. Single terminal contractile vacuole. Division takes place in thin-walled cysts, thick-walled protective cysts also formed.

Could be easily confused with *Wenrichia* (see Hymenostomata Part II).

Keys to species. Hukui (1956), Kahl (1930–35).

Description of species. Burt (1940).

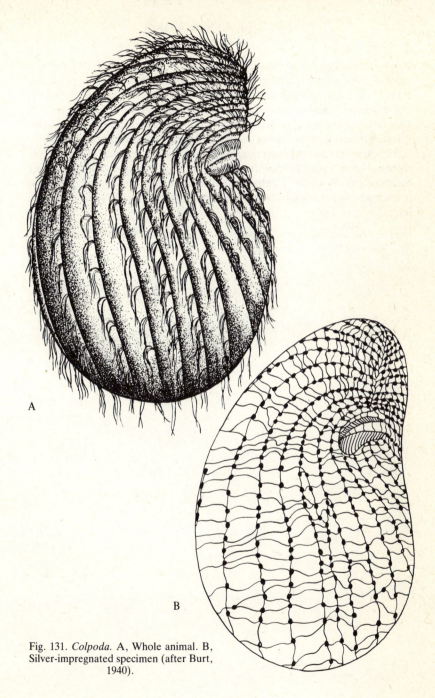

A

B

Fig. 131. *Colpoda.* A, Whole animal. B, Silver-impregnated specimen (after Burt, 1940).

Cyrtolophosis Stokes, 1885
(Fig. 132)

Description. Shape ovoid to pyriform, small (20–40 μm long). Body uniformly ciliated, cilia in pairs arranged in kineties which slightly curve down the body. There is an anterior tuft of cilia in certain species. The oral aperture is in a prominent shallow groove which occupies the anterior third of the cell. Oral ciliature consists of a haplokinety on the right (paroral ciliature) in two segments and four small adoral organelles forming a membranelle on the left. Some species construct gelatinous tubes which they may leave. Macronucleus spherical and centrally positioned, micronucleus enclosed within outer macronuclear envelope. Contractile vacuole situated in posterior body third connected to a permanent, posteriorly directed tubular canal several microns long.

Descriptions and taxonomy. Detcheva (1976), Foissner (1978a), Kahl (1930–35), McCoy (1974), Stokes (1885b).

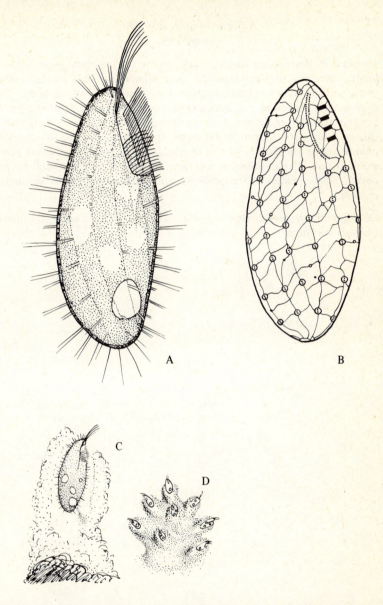

Fig. 132. *Cyrtolophosis*. A, Whole animal. B, Silver-impregnated specimen (both after Foissner, 1978a). C, Animal in housing. D, Several housings clustered together (both drawn from life).

Kalometopia Bramy, 1962
(Fig. 133)

Description. Body shape approximately oval in outline, quite large (250 μm long). There is a well-developed anterior ciliated oral field on the ventral surface which occupies the majority of the anterior half of the body. The depression is shallow at the edges and deepens towards the cytostome which lies approximately in the middle of the ventral surface. The body is uniformly ciliated with many longitudinal/oblique rows which originate at a curved preoral suture on the left of the vestibular opening. Division takes place within a cyst. Macronucleus in several parts, and there are several micronuclei. Single posterior contractile vacuole situated on left of vestibular opening.

Descriptions of species. Bramy (1962), Gellért (1950b – under the name *Colpoda eurystoma*).

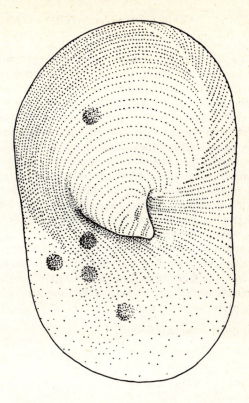

Fig. 133. *Kalometopia* (after Bramy, 1962).

Kuklikophrya Njiné, 1979
(Fig. 134)

Description. Outline shape approximately oval with slight anterior rounded rostrum that bends somewhat to the left. The posterior body region is narrower than the anterior and terminates in a blunt point. The oral aperture lies in the centre of the anterior third of the ventral surface in the form of a diagonal slit. On the right of the oral aperture there are two parallel rows of kinetosomes which travel posteriorly to encircle the mouth and extend up its left side. This circumoral ciliature meets a row of preoral membranelles which are composed of 10–12 blocks of cilia that are arranged in paired rows. The preoral ciliature begins on the left side of the anterior third of the oral aperture and passes forward along the rostrum. The body is uniformly ciliated and the cilia arise in pairs. The centrally placed macronucleus is spherical and there is a single posterior contractile vacuole. Single record from temporary pools in Africa.

Most easily mistaken for *Enigmostoma* (p. 244) which has a mouth supported by short trichites but does not have circumoral ciliature. It may be distinguished from the related genus *Rostrophrya* (p. 208) by the position of the preoral membranelles; in the latter genus the line of membranelles extends to the posterior of the oral aperture.

Description of species. Njiné (1979a).

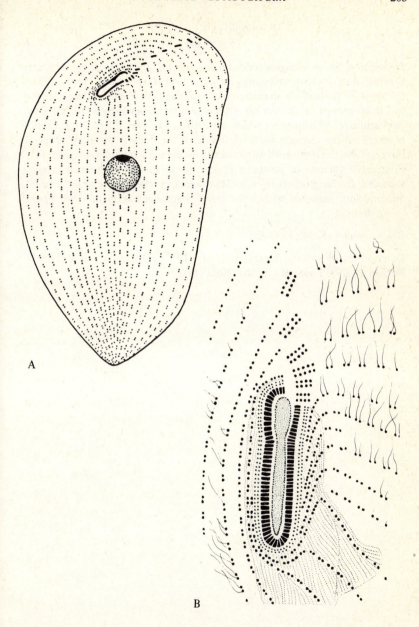

Fig. 134. *Kuklikophrya*, silver-impregnated specimens. A, Ventral surface. B, Oral area (after Njiné, 1979a).

Platyophrya Kahl, 1926
(Fig. 135)

Description. Shape elongate reniform to pyriform, highly metabolic with very short left anterior rostrum bearing an apical oval oral aperture. On the right of the aperture there is a C-shaped row of paired cilia (paroral ciliature) and five to fifteen groups of six cilia on the left (adoral ciliary membranelle). Body cilia uniform, cilia in pairs which arise from furrows. Silver impregnation reveals that the ciliation on the left side of the body is distinctly reduced, this results in the meshwork infraciliature on the right and left sides (Fig. 135). Contractile vacuole in posterior region, some species with small satellite vacuoles. Division occurs in cysts. Macronucleus rounded, centrally located, micronucleus included within the outer macronuclear membrane. Found living in moss.

Key to species. Kahl (1930–35).

Descriptions of species. Dragesco *et al.* (1977), Dragesco and Dragesco–Kerneis (1979), Foissner (1978a), Kahl (1926).

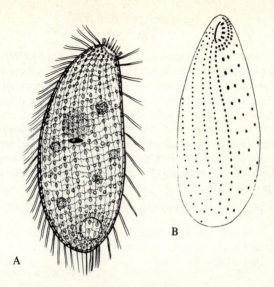

Fig. 135. *Platyophrya*. A, Lateral view. B, Ventral view (silver impregnations after Foissner, 1978a).

Puytoraciella Njiné, 1979
(Fig. 136)

Description. Body outline shape oval, uniformly ciliated by paired cilia. Oval oral aperture in anterior body third. On the right of the aperture there is a paroral membranelle arising from kinetosomes that are arranged in a zig-zag manner. The paroral membrane travels down, around the posterior end of the oral aperture and halfway up its left side. A line of membranelles begins at the base of the aperture and this extends obliquely left onto the dorsal body surface. Each membranelle arises from a block of kinetosomes that is made up from three rows which is similar to the arrangement found in *Bryophrya* (p. 192). Cytoplasm yellowish, feeds on Cyanophyceae, found in temporary pools in Africa.

Description of species. Njiné (1979b).

Taxonomy. Puytorac, Perez-Paniagua and Perez-Silva (1979).

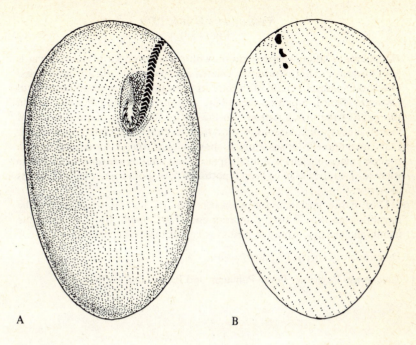

Fig. 136. *Puytoraciella*, silver impregnated specimens. A, Ventral aspect. B, Dorsal aspect (after Njiné, 1979b).

Rostrophrya Njiné, 1979
(Fig. 137)

Description. Outline shape reniform with pronounced short rostrum that is bent towards the left. Uniformly ciliated all over body, cilia arising in pairs. Oval oral aperture situated in centre of ventral surface near to the base of rostrum. There is a paroral membrane with paired ciliary bases situated on the right of the oral aperture. On the left of the mouth there is a line of membranelles which extends forward from its base to the end of the rostrum. Each membranelle arises from a short double row of kinetosomes. The rounded macronucleus is centrally placed and there is a terminal contractile vacuole. Two species have been reported to date, both from temporary pools in Africa.

Most easily mistaken for *Kuklikophrya* (p. 202) in which genus the membranelles arise from the anterior end of the mouth. In *Rostrophrya* the membranelles arise from the posterior end of the oral aperture.

Description of species. Njiné (1979b).

Taxonomy. Puytorac, Perez-Paniagua and Perez-Silva (1979).

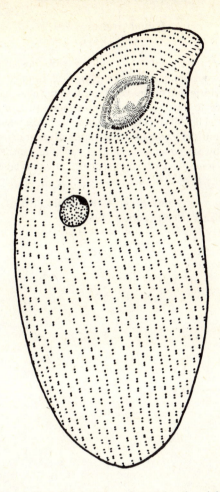

Fig. 137. *Rostrophrya.* Silver impregnation of ventral surface (after Njiné, 1979b).

Tillina Gruber, 1879
(Fig. 138)

Description. Body approximately reniform with posterior end twisted. Somatic groove winds obliquely around body back from entrance of vestibulum. Ciliation uniform and complete in spiral rows with cilia often emerging in pairs. Entrance to vestibulum on ventral body surface with prominent semicircular ventral lip which bears a single row of closely set cilia on the left border. Vestibulum deep. Single large terminal contractile vacuole which may be served by six to nine very long radiating canals. Macronucleus ellipsoidal, many micronuclei (two to sixteen). Division takes place within a thin-walled cyst.

Descriptions of species. Fauré-Fremiet and André (1965a), Gruber (1880), Lynn (1976), Suhama (1969), Turner (1937).

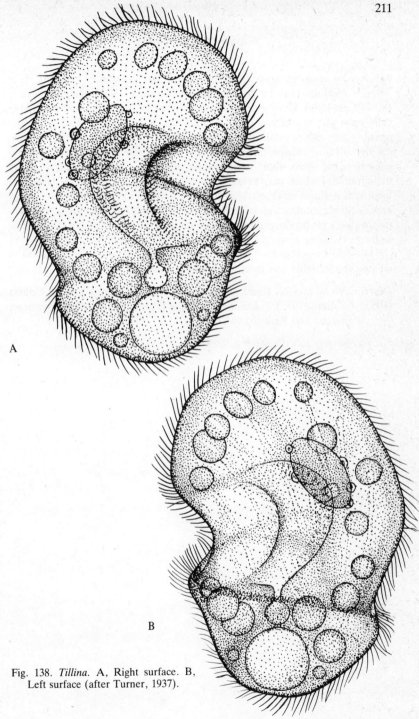

Fig. 138. *Tillina*. A, Right surface. B,
Left surface (after Turner, 1937).

Woodruffia Kahl, 1931
(Fig. 139)

Description. Body elongate reniform to pyriform, sometimes with very short anterior rostrum, may be dorso-ventrally flattened. Highly variable in size (80–400 µm long). Body entirely covered with cilia, uniformly and symmetrically arranged in oblique rows. Cilia often arise in pairs. The vestibular opening may either be slit-like or oval in the anterior region of the cell. On the right of the oral aperture there is a C-shaped row of paired cilia (paroral ciliature) and many short double rows of cilia on the left (adoral ciliary membranelle) which may extend forwards slightly. Macronucleus large and oval with a single micronucleus contained within the macronuclear membrane. Single contractile vacuole terminal. Division takes place within a thin-walled cyst producing two to four daughters. Resistant cysts with thick walls may also be formed.

Most easily confused with *Platyophrya* (p. 204) which has a reduction of cilia on the left side and hence is unsymmetrical.

Descriptions of species. Beers and Sherwood (1966), Czapik (1971), Gellért (1955), Grolière (1975), Johnson and Larson (1938), Prelle (1963), Puytorac, Perez-Paniagua and Perez-Silva (1979)

Taxonomic revision. Foissner (1978a).

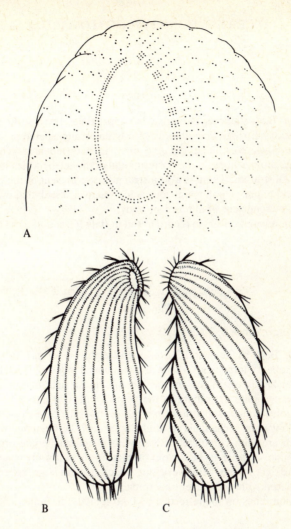

Fig. 139. *Woodruffia*. A, Silver impregnation of oral area (after Puytorac *et al.*, 1979). B, C, Ventral and dorsal surfaces respectively (after Czapik, 1971).

INCERTAE SEDIS TRICHOSTOMATIDA

Grandoria Corliss, 1960
(Fig. 140)

Lagenella Grandori and Grandori, 1934

Description. Body shape asymmetrical but approximately pyriform in outline with posterior narrower than anterior. Body uniformly ciliated in longitudinal/oblique rows. Vestibular opening oval, in body equator. There is a bundle of gelatinous fibres at the posterior end of the cell by which it attaches itself to a substratum. There is also a strong bristle present at the posterior which is held to one side. Several contractile vacuoles distributed over body. Spherical macronucleus in anterior of body.

Description based on single record in soil in Italy (Grandori and Grandori, 1934).

Opisthostomatella Corliss, 1960
(Fig. 141)

Opistostomum Ghosh, 1928

Description. Body shape ovoid. Posterior end slightly tapering. Anterior end with three lobes, a large ventral lobe and two dorsal lobes of which the right is the larger and overlaps the ventral lobe along the right margin. A deep furrow (the vestibulum?) separates the ventral and right dorsal lobes. The small left dorsal lobe is separated from the ventral lobe by a shallow notch. There is a long narrow channel (vestibular aperture?) between the two dorsal lobes which are both completely edged with membranelle-like organelles. The vestibulum appears to be devoid of cilia. Body uniformly ciliated with those on the right margin being very long. Macronucleus large, spherical and centrally placed with an adjacent micronucleus.

Description relies on single record by Ghosh (1928) in India.

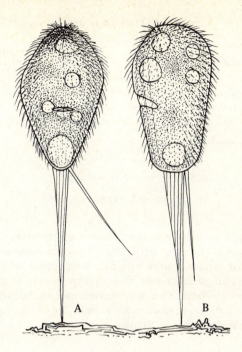

Fig. 140. *Grandoria.* A, Ventral aspect. B, Lateral aspect (after Grandori and Grandori, 1935).

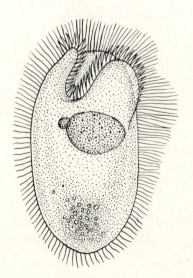

Fig. 141. *Opisthostomatella* (after Ghosh, 1928).

Orcavia Tucolesco, 1962
(Fig. 142)

Description. Shape highly symmetrical, oblong in outline with rounded ends. Round in cross-section. Uniformly ciliated in 15–18 ciliary rows on each side. There is an oval apical vestibular aperture opening into a large oblong vestibulum which contains a fine sinusoidal flap or flange. Macronucleus in several parts. Found on one occasion in a stagnant pool in a cave (Romania).

Description. Tucolesco (1962).

Rigchostoma Vuxanovici, 1963
(Fig. 143)

Description. Shape reniform with short rostrum which curves to the right (in other genera it curves to the left). Ventral surface flattened with 16–18 striations. Dorsal surface convex with four longitudinal ridges. Body uniformly ciliated in longitudinal rows. Oval oral aperture in an apical position at end of rostrum and bears a triangular membrane. Macronucleus spherical in anterior body half. Contractile vacuole centrally placed. Based on single record by Vuxanovici (1963) in Romania.

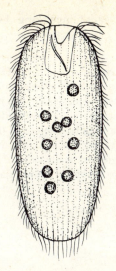

Fig. 142. *Orcavia* (after Tucolesco, 1962).

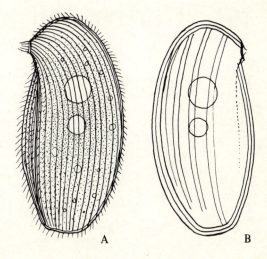

Fig. 143. *Rigchostoma*. A, Left side. B, Right side (after Grandori and Grandori, 1935).

Sulcigera Gajewskaja, 1928
(Fig. 144)

Description. Shape circular in outline, dorso-ventrally flattened but the dorsal side is more arched than ventral surface. On the ventral surface there is a deep groove which originates at the oval subapical oral aperture and ends terminally. The groove is lined on each side with long cilia. At the posterior end of the groove are four cirri and on either side of the oral aperture is a semi-circular ridge bearing long cilia. The rest of the body is covered in long cilia which tend to be longer on the left side. Macronucleus elongate to ovoid, centrally placed. Single contractile vacuole in posterior region of cell.

Descriptions based on two reports (Gajewskaja, 1928, 1933) from plankton in Lake Baikal.

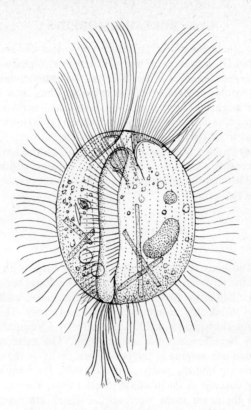

Fig. 144. *Sulcigera* (after Gajewskaja, 1928).

Order BURSARIOMORPHIDA

Until recently the members of this order were included in the subclass Spirotricha (Part II), however, the work of Fernandez-Galiano (1978) on the structure of the membranelles of *Bursaria* has shown that they are more like those of the 'cirromembranelles' of the colpodids than those of the spirotrichs. Similar work by Gerassimova, Sergejeva and Seravin (1979) is in general agreement with that of Fernandez-Galiano (1978) even though these workers appear to be ignorant of the latter author.

The bursariomorphids are large freshwater ciliates which have a truncated anterior end bearing a large deep funnel-like peristome region. The membranelles wind around the apical region of the peristome and plunge down the left side of the funnel.

Bursaria Müller, 1773
(Fig. 145)

Description. The body is approximately ovoid in shape with the anterior truncated and the posterior broadly truncated. The dorsal surface is convex and the anterior half of the flattened ventral surface is divided by a wide slot which leads to the deeply invaginated peristomial cavity. The peristomial cavity opens anteriorly and is divided internally by a longitudinal fold; the cytopharynx is bent towards the animal's left. The membranelles wind clockwise around the opening of the peristomial cavity on the anterior edge of the body and dip into the cavity on the left side. The body is completely and uniformly covered in longitudinal rows of cilia. The macronucleus is band-like and there are many micronuclei. There are many contractile vacuoles which are distributed along the lateral and posterior edges of the body.

Most easily confused with *Thylakidium* (p. 223), *Bursaridium* (p. 222) and *Climacostomum* (Part II). Unlike *Bursaria*, the latter genus (a true spirotrich) does not have a deep anterior peristomial funnel. In *Bursaridium* (Fig. 120) the cytopharynx bends to the right while in *Bursaria* and *Thylakidium* it bends to the left. *Bursaria* is distinguished from *Thylakidium* by the former's possession of a longitudinal fold in the peristomial cavity.

Descriptions. Fernandez-Galiano (1978), Gerassimova, Sergejeva and Seravin (1979).

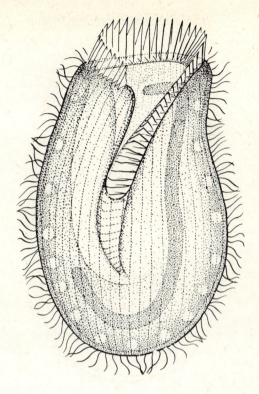

Fig. 145. *Bursaria* (composite from many sources).

Bursaridium Lauterborn, 1894

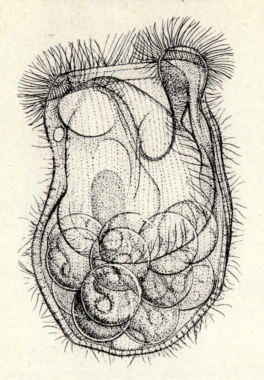

Fig. 146. *Bursaridium* (from life).

Description. Body approximately ovoid in shape with the anterior markedly truncate and the posterior broadly rounded. The ventral surface is divided by a wide slot which leads into the deeply invaginated peristomial cavity which opens apically. The cytopharynx is bent towards the animal's right (Fig. 120) which distinguishes this genus from *Bursaria* (p. 220) and *Thylakidium* (p. 223) in which genera the cytopharynx bends towards the animals' left. There is a line of membranelles which winds clockwise around the opening of the peristomial cavity on the anterior edge of the body and dips down into the cavity on the left. The body is uniformly ciliated. The macronucleus is elongate and there is a single terminal contractile vacuole.

Description. Fauré-Fremiet (1924).

Thylakidium Schewiakoff, 1892

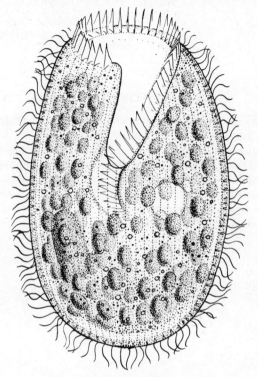

Fig. 147. *Thylakidium* (composite from several sources).

Description. Body ovoid with anterior truncated and posterior broadly rounded. The ventral surface is divided by a wide slot which leads to the deep peristomial cavity which opens apically. The cytopharynx (Fig. 120) is bent towards the animal's left. There is a line of membranelles which winds clockwise around the apex of the cell and plunges into the peristomial funnel on the left. The body is uniformly ciliated. The macronucleus is rounded and there is a single laterally positioned contractile vacuole. The cytoplasm is usually full of green zoochlorellae.

Most easily confused with *Bursaria* (p. 220) which, unlike *Thylakidium*, has a divided peristomial cavity (Fig. 120). It could also be mistaken for *Bursaridium* (p. 222) whose cytopharynx bends to the right not to the left as in *Thylakidium*.

Keys and descriptions of genera of Hypostomata

Subclass HYPOSTOMATA

In the hypostomes the oral apparatus is situated on the ventral surface of the body which may either be cylindrical or dorso-ventrally flattened. The somatic ciliation is often reduced and the cytopharyngeal apparatus is of the cyrtos type. The oral area may be sunk into an atrium with organised atrial ciliature present. Morphogenesis is often complicated, with stomatogenesis of an advanced telokinetal type or even parakinetal- or buccokinetal-like. Toxicysts are generally absent but trichocysts are present in one group. Many species are endo- or ectosymbionts usually growing in or on invertebrate hosts.

Key to genera of free-swimming Hypostomata

1. Intimately associated with, and feeding upon, the exuvial fluid of the moult of crustaceans **32**

   Not associated with crustaceans **2**

2. Cell with one or more posterior protoplasmic spines ....
   .... *See sessile Hypostomata* (p. 232)

   Cell without posterior spine **3**

3. Always with post-oral frange (Fig. 148A,B) which usually extends to both right and left of oral aperture, often over onto dorsal surface. When the frange does not extend past the right of the aperture, the cilia of the frange arise from a single row of simple kinetosomes. Always ciliated on all body surfaces **4**

   Post-oral frange not always present, but when it is the cilia arise from compound kinetosomes which do not extend past right of oral aperture (Fig. 148C), in some the frange is reduced to not easily observable pseudomembranelles (Fig. 148D). Not always ciliated on dorsal surface **7**

4. Dorso-ventrally flattened **5**

   Rounded in cross-section **6**

5. Anterior end distinctly narrowed and partly elongated to form a rounded proboscis which is held towards the left ..... *Orthodonella* (p. 238)

   Anterior end broadly rounded, never narrows but may be bent towards the left .... *Chilodontopsis* (p. 234)

6. With definite beak directed to the left .... *Synhymenia* (p. 240)

   Without beak. Frange born upon low ridge .... *Nassulopsis* (p. 236)

7. Body compressed laterally, always encased in rigid pellicle which may be ribbed. Cells tend to be small (under 30 μm long). Commonly two contractile vacuoles present but some have one **13**

   Body may be compressed dorso-ventrally but is never encased in a rigid pellicle. Cells tend to be larger (over 30 μm long). Commonly with single contractile vacuole but some have two **8**

8. Body entirely ciliated, usually rounded in cross-section **9**

   Body ciliated only on ventral surface, usually dorso-ventrally flattened, but two genera are not **23**

9. Contractile vacuole never in front of oral aperture **10**

   One of the two contractile vacuoles lies in front of the oral aperture .... *Chilodina* (p. 242)

10. Body smooth, without transverse/oblique ridges **11**

    Body covered in many low transverse/oblique ridges ....

    .... *Archinassula* (p. 242)

A Chatton-Lwoff silver preparation is necessary for next two stages

11. Pseudomembranelles lie along suture line and do not interupt kineties on left of body (Fig. 148D) **12**

    Pseudomembranelles lie across kineties situated on left side of body not along suture line (Fig. 148C) .... *Nassula* (p. 248)

12. Only three membranelles present (Fig. 148D) .... *Furgasonia* (p. 246)

    Many (approximately ten) membranelles present ....

    .... *Enigmostoma* (p. 244)

13. With cytopharyngeal basket of trichites **14**

    Without cytopharyngeal basket of trichites **17**

14. Oral aperture mounted on anterior protuberance of pellicle (Fig. 148E) .... *Stammeridium* (p. 262)

    Oral aperture not mounted on protuberance **15**

15. With two contractile vacuoles .... *Drepanomonas* (p. 250)

    With single contractile vacuole **16**

A Chatton-Lwoff silver preparation is necessary for next stage

16. With three membranelles on left of oral aperture. Approximately fifteen somatic kineties .... *Pseudomicrothorax* (p. 260)

    Without three membranelles on left of oral aperture. Approximately eight somatic kineties .... *Leptopharynx* (p. 256)

17. With two contractile vacuoles **18**

    With single contractile vacuole **19**

18. Opening of oral aperture in posterior region of body **22**

    Opening of oral aperture in centre of body edge .. *Drepanomonas* (p. 250)

19. Opening of oral aperture in posterior half of body ..... *Kreyella* (p. 254)

    Opening of oral aperture in anterior half of body **20**

20. With anterior lip or beak **21**

    Anterior end rounded, without lip or beak ....

    .... *Microdiaphanosoma* (p. 258)

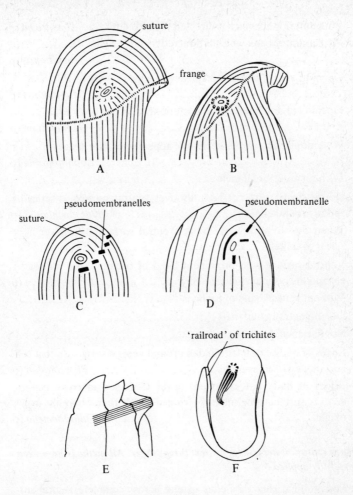

Fig. 148. Oral structures in some hypostome ciliates. A, *Synhymenia* (after Jankowski, 1968). B, *Orthodonella* (after Jankowski, 1968). C, *Nassula* (after Fauré-Fremiet, 1959). D, *Furgasonia* (after Fauré-Fremiet, 1967b). E, *Stammeridium* (after Wenzel, 1969). F, *Chlamydodon* (after Kaneda, 1953).

21. With anterior lip which is directed towards left ...... *Hexotricha* (p. 252)
With anterior beak which is directed towards right ....
.... *Trochiliopsis* (p. 264)

22. Opening of oral aperture on posterior edge of body ....
.... *Microthorax* (p. 260)
Opening of oral aperture on posterior half of left body edge ......
.... *Hemicyclium* (p. 252)

23. With mobile anterior tentacle-like appendage (rare genus) ....
.... *Lophophorina* (p. 272)
Without anterior tentacle **24**

24. Dorsal surface extends laterally over ventral surface as series of
spikes around edges .... *Odontochlamys* (p. 276)
When dorsal surface overhangs ventral surface it is never as a
series of spikes **25**

25. Ventral surface with prominent band of trichites (the 'railroad')
around the periphery (Fig. 148F) .... *Chlamydodon* (p. 272)
Without ventral band of trichites **26**

26. Dorso-ventrally flattened **28**
Not flattened dorso-ventrally **27**

27. Edges of body extended, folded around ventral surface so that cell
is C-shaped in cross-section .... *Phascolodon* (p. 278)
Edges of body not extended, never C-shaped in cross-section,
ventral surface longitudinally ridged, with four hoop-like rows
of cilia .... *Chilodonatella* (p. 266)

*Silver preparation necessary for next three stages. All methods have been
successfully applied*

28. Longitudinal ciliary kineties extend across complete ventral surface, there is never a central gap (Fig. 149A,B) **29**

Longitudinal ciliary kineties do not completely extend across
ventral surface, there is always a central gap in the rows (Fig.
149C–E) **30**

29. Pre-oral kinety Y-shaped, without a kinety along suture line
(Fig. 149A) .... *Chlamydonella* (p. 274)

Pre-oral kinety never Y-shaped, with a kinety along suture line
(Fig. 149B) .... *Trithigmostoma* (p. 280)

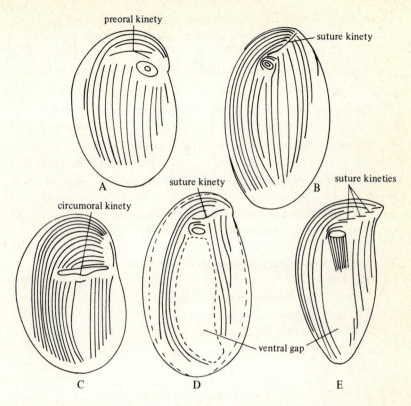

Fig. 149. Ventral ciliary patterns of some hypostome ciliates. A, *Chlamydonella* (after Patsch, 1974). B, *Trithigmostoma* (after Jankowski, 1967b). C, *Gastronauta* (after Wilbert, 1972). D, *Chilodonella* (after Jankowski, 1967b). E, *Pseudochilodonopsis* (after Foissner, 1979).

30. A kinety completely surrounds oral aperture which is not supported by trichites. The oral aperture is slit-like and segregates left and right fields of kineties (Fig. 149C) .... *Gastronauta* (p. 270)

A kinety never surrounds the oral aperture which is supported by a basket of trichites **31**

31. There is a single long kinety along suture line (Fig. 149D) .... *Chilodonella* (p. 268)

There are several short kineties along suture line (Fig. 149E) .... *Pseudochilodonopsis* (p. 282)

Chatton-Lwoff or Protargol silver preparation necessary for next two stages

32. Kinety nine begins in middle of body (Fig. 150B,C) **33**

Kinety nine begins at apex of body (Fig. 150A) .... *Gymnodinioides* (p. 294)

33. With anterior ventral field composed of groups of two or three cilia (Fig. 150C) .... *Hyalophysa* (p. 296)

Without an anterior ventral field of cilia (Fig. 150B) .... *Hyalospira* (p. 296)

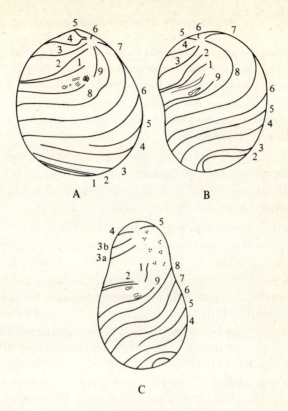

Fig. 150. Ciliary meridians in apostome ciliates. A, *Gymnodinioides* (after Chatton and Lwoff, 1935). B, *Hyalospira* (after Miyashita, 1933). C, *Hyalophysa* (after Grimes, 1976).

Key to genera of sessile Hypostomata

1. Outline shape oval, dorso-ventrally flattened. Attached to substratum by fine thread which is secreted from one or two posterior
cytoplasmic spines. Without anterior funnel. (Cyrtophorida) **2**

   Elongate, rounded in cross-section. Attached by posterior end of
body to a crustacean. No spines present. Apical region opens into a
spiral funnel. (Chonotrichida) **6**

2. With extensive left field of kineties which extend continuously
across the ventral surface beneath the oral aperture (Fig. 151A–C)   **3**

   Left field of kineties reduced and is in two separate areas, one lies
anteriorly, the other centrally (Fig. 151D,E) **4**

3. Oral aperture supported by many trichites (Fig. 151B,C) **5**

   Oral aperture supported by only six trichites which are armed with
teeth (Fig. 151A) .... *Trochilioides* (p. 288)

4. Anterior part of left ciliary field lies in front of oral aperture which
is supported by two wide blunt elements; four or more kineties to
the right field (Fig. 151E) .... *Dysteria* (p. 284)

   Anterior part of left ciliary field lies to left of oral aperture which is
supported by two trichites each armed with needle-like teeth;
always four kineties to right field (Fig. 151D) .... *Trochilia* (p. 288)

5. The two short kineties in front of the oral aperture lie transversely
end to end (Fig. 151B) .... *Orthotrochilia* (p. 284)

   The two short kineties in front of the oral aperture lie transversely
side by side (Fig. 151C) .... *Parachilodonella* (p. 286)

6. Rim of outer whorl of pre-oral funnel is even and without notches.
Tail region short. Common on freshwater gammarid crustaceans in
all of Europe .... *Spirochona* (p. 292)

   Rim of outer whorl of pre-oral funnel interrupted by one or more
notches. Tail region elongate. Only reported on crustaceans from
Lake Baikal **7**

7. Single notch in outer whorl of a well-developed pre-oral funnel
.... *Cavichona* (p. 290)

   Several notches (up to eight) in rim of outer whorl of under-
developed pre-oral funnel .... *Serpentichona* (p. 292)

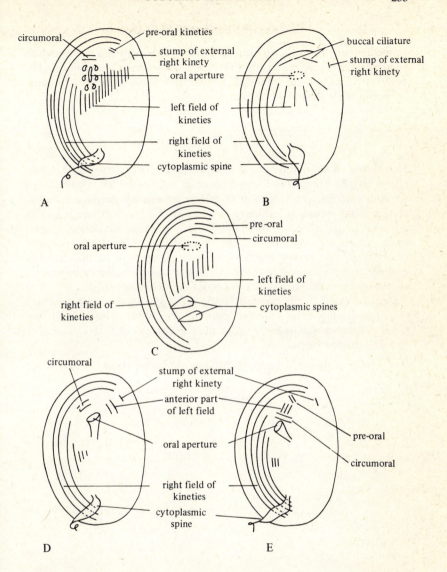

Fig. 151. Comparison of ventral ciliary patterns in Dysteriina. A, *Trochilioides* (after Deroux, 1976). B, *Orthotrochilia* (after Patsch, 1974). C, *Parachilodonella* (after Wilbert, 1972). D, *Trochilia* (after Heuss and Wilbert, 1973). E, *Dysteria* (after Fauré-Fremiet, 1965).

Order SYNHYMENIIDA

In members of this order there is an extensive hypostomial frange which winds around the anterior part of the body which is usually cylindrical. The body is covered by holotrichous ciliation and the cilia usually emerge in pairs. Stomatogenesis is parakinetal-like.

Chilodontopsis Blochmann, 1895
(Fig. 152)

Description. Body outline shape oval, dorso-ventrally flattened. Anterior region broadly rounded and slightly directed towards the left. Body ciliated on both surfaces. Oral aperture in middle of anterior quarter of ventral surface. Oral apparatus consisting of a basket of trichites. There is a short post-oral frange composed of an oblique line of simple but distinct kinetosomes which does not traverse the complete ventral surface, instead it reaches only one or two kineties past the oral aperture on the right but extends to the left round to the dorsal surface. Single posterior contractile vacuole. Macronucleus more or less ovoid, never sausage-shape.

Genus created by Blochmann (1895) for a species described by Perty (1852) under the name *Chilodon depressa*. Jankowski (1968) regards the genus as a *nomen nudum* but in the opinion of Deroux (1978) this is a distinct and well-defined genus.

Descriptions of species. Deroux (1978), Fauré-Fremiet (1959, 1967a).

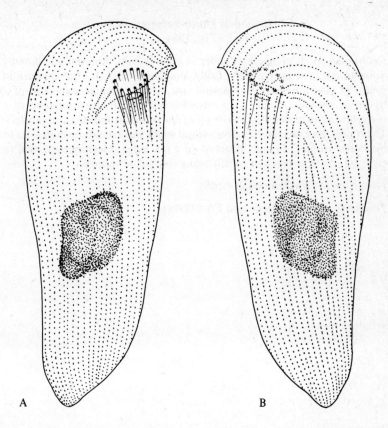

A   B

Fig. 152. *Chilodontopsis*. A, Ventral surface. B, Dorsal surface (after Deroux, 1978).

Nassulopsis Fauré-Fremiet, 1959
(Fig. 153)

Description. Body elongate, rounded in cross-section. Anterior end broadly rounded not bent towards left. Body completely ciliated. Oral aperture in centre of anterior quarter of ventral side. Oral apparatus composed of a basket of trichites. There is an extensive post-oral frange composed of a transversely orientated (slightly oblique) line of compound ciliary organelles. The frange traverses the complete ventral width and extends on both sides to the dorsal surface. Frange mounted on a low ridge. Macronucleus oval to elongate. At least two and usually many contractile vacuoles.

List of species. Fauré-Fremiet (1959).

Taxonomy. Deroux, Iftode and Fryd (1974).

Fig. 153. *Nassulopsis* (after Deroux, Iftode and Fryd, 1974).

Orthodonella Bhatia, 1936
(Fig. 154)

Orthodon Gruber, 1884

Description. Body elongate with anterior distinctly narrowed to form a blunt pointed 'beak' which is strongly bent towards the left. Body dorso-ventrally flattened and ciliated on both surfaces. Oral aperture in anterior body quarter but displaced to the right. Oral apparatus supported by a basket of trichites. There is a distinct obliquely orientated post-oral frange composed of a line of compound ciliary organelles. The frange extends left anteriorly up the pointed beak region and posteriorly past the oral aperture almost to the right body edge. There is a large, usually terminal, contractile vacuole and the macronucleus is ovoid.

Descriptions of species. Gruber (1884), Jankowski (1968).

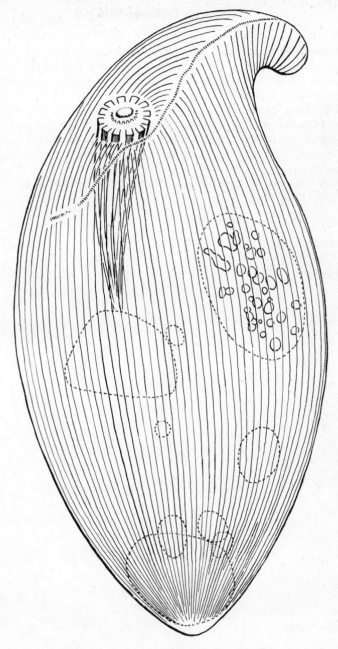

Fig. 154. *Orthodonella* (after Jankowski, 1968).

Synhymenia Jankowski, 1967
(Fig. 155)

Description. Body elongate, rounded in cross-section with the anterior quarter bent strongly to the left in the form of a beak. Body ciliated completely. Oral aperture in centre of anterior quarter of ventral surface. Oral apparatus consists of a basket of trichites. There is an extensive post-oral frange composed of a transversely orientated line of compound ciliary organelles which traverses the complete ventral width and extends on both sides to the dorsal surface. Several small vacuoles distributed throughout the cell plus a single large terminal contractile vacuole present. Macronucleus ovoid.

Genus created by Jankowski (1967c) for the species *Nassula heterovesiculata* Gelei, 1939).

Description of species. Gelei (1939).

Taxonomy. Jankowski (1968).

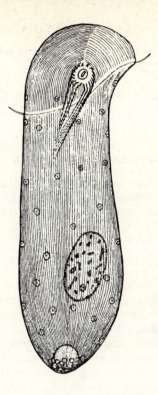

Fig. 155. *Synhymenia* (after Gelei, 1939 and Jankowski, 1968).

Order NASSULIDA

Here the hypostomial frange is not often extensive and is either limited to the left side of the ventral surface or is sometimes reduced to three or four 'pseudomembranelles' in an oral atrium. There is a distinct pre-oral suture present. Stomatogenesis is complicated, being parakinetal- or buccokinetal-like in many species. Trichocysts are characteristically present in one suborder.

Suborder NASSULINA

These genera have the characteristics of the above order. Typically the body is pliable, large and cylindrical, covered in a coat of holotrichous cilia. There are both marine and freshwater species which typically feed upon filamentous algae.

Archinassula Kahl, 1935
(Fig. 156)

Description. Outline shape, slim ovoid, slightly curved to the right. Body striated obliquely by low but sharp ribs which encircle the body. Somatic cilia arise between the ribs and ciliature of body appears to be complete. The oral aperture lies in a shallow groove in the extreme anterior region and is supported by a cytopharyngeal basket of trichites. There appears to be no adoral ciliature. Macronucleus large, ovoid to kidney shape, centrally positioned with an adjacent micronucleus. Single equatorial contractile vacuole situated on right of body.

Genus based on single brief description by Kahl (1930–35), Germany.

Chilodina Šrámek-Hušek, 1957
(Fig. 157)

Description. Body outline shape oval, rounded in cross-section. Completely ciliated on all surfaces of body. Oral aperture in anterior third of body in centre of ventral surface, supported by cytopharyngeal basket of trichites. Two contractile vacuoles present, one which lies in front of the oral aperture has radiating canals. The second is situated terminally. Macronucleus ovoid, centrally placed.

Description of species. Šrámek-Hušek (1957).

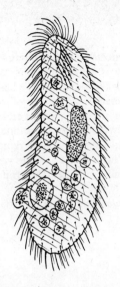

Fig. 156. *Archinassula* (after Kahl, 1930–35).

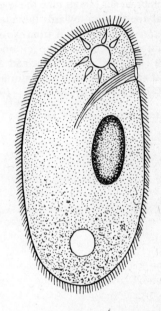

Fig. 157. *Chilodina* (after Šrámek-Hušek, 1957).

Enigmostoma Jankowski, 1975
(Fig. 158)

Description. Body outline shape oval to reniform with an anterior beak directed left, highly metabolic. Body ciliation complete, in pairs forming many longitudinal rows. The oral aperture, which is supported by a basket of trichites, is slit-like and longitudinally orientated. A frange of several membranelles (about nine to eleven) leads from the anteriormost end of the oral aperture up towards the apex of the body along the suture line. This feature distinguishes it from *Nassula* (p. 248). There is a large equatorial contractile vacuole on the left body edge and two smaller satellite vesicles, one equatorial on the right and a posterior one on the left. Macronucleus oval and approximately equatorial.

Genus erected by Jankowski (1975) for *Nassula ougandae* Dragesco, 1972.

Description of species. Dragesco (1972).

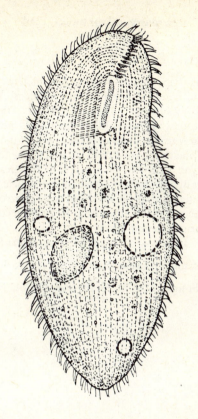

Fig. 158. *Enigmostoma* (after Dragesco, 1972).

Furgasonia Jankowski, 1964
(Fig. 159)

Cyclogramma Perty, 1852

Description. Outline shape oval to elongate. Completely ciliated (sometimes with paired cilia) in about 35 longitudinal meridians. Without frange of membranelles as in *Nassula* (p. 248) but is reduced to three membranelles lying on left of the oral aperture. Membranelles do not interupt the left ventral kineties as in *Nassula* but occur in the region of the suture line. There is a double row of cilia to the right of the oral aperture forming the paroral ciliature. Cytopharynx supported by a basket of trichites. Fusiform trichocysts always present in large numbers beneath the pellicle. Single contractile vacuole always situated ventrally in middle of body. Macronucleus large with adjacent micronucleus centrally positioned. Body often coloured blue-green to red due to colour of ingested algae.

Genus erected by Jankowski (1964) for *Cyclogramma* Perty, 1852.

Description of species and revision. Fauré-Fremiet (1967b)

Taxonomy. Fauré-Fremiet (1967a), Grain *et al.* (1976).

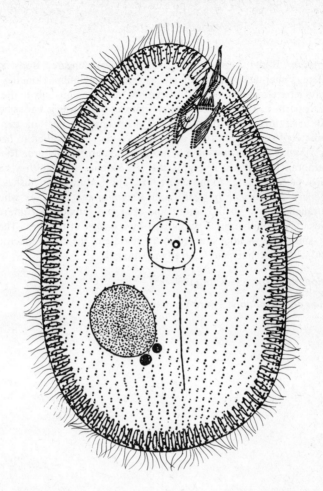

Fig. 159. *Furgasonia* (composite from Fauré-Fremiet, 1976b).

Nassula Ehrenberg, 1833
(Fig. 160)

Description. Body shape ovoid, some species elongate. Body ciliature complete, sometimes with paired cilia, in many longitudinal kineties. There is always a frange of membranelles beneath, but extending only to the left of, the oral aperture. The number of membranelles in the frange varies but there are at least three and usually more. The anterior part of the kinety immediately to the right of the oral aperture may be modified to become a highly vibratile double row of cilia forming the paroral ciliature. The frange of membranelles always interupts the left kineties so that they continue forwards past the frange to the suture line. This feature distinguishes it from the genera *Furgasonia* (p. 246) and *Enigmostoma* (p. 244) where the membranelles occur in the suture line area and thus do not interupt the kineties. The cytopharynx is supported by a basket of trichites. There is a single contractile vacuole situated in the middle of the body. Macronucleus with adjacent micronucleus centrally positioned.

Descriptions of species. Kahl (1930–35).

Revision and taxonomy. Fauré-Fremiet (1959, 1967a).

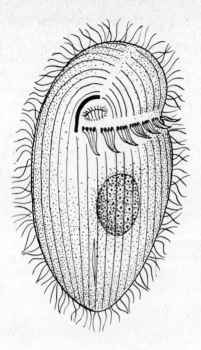

Fig. 160. *Nassula* (after Corliss, 1979).

Suborder MICROTHORACINA

In this suborder the hypostomial frange is drastically reduced to a few kinetal segments bearing 'pseudomembranelles' which are sometimes set in a shallow atrium. The body is rigid, usually broadly ellipsoidal or crescent-shaped and is often laterally flattened. Ciliation on the body tends to be sparse but fibrous trichocysts are characteristically present.

Drepanomonas Fresenius, 1858
(Fig. 161)

Description. Outline shape sometimes oval, sometimes like a segment of an orange with both anterior and posterior ends pointed. Flattened with rigid ribbed armour, some species with spines. Somatic ciliature reduced with fewer on dorsal surface than on ventral surface. Opening to cytostome is in a depression in the left side, located in centre of the body. There is sometimes a pharyngeal basket of trichites and a membranelle on the anterior left of the oral aperture. There is a centrally placed macronucleus and two central contractile vacuoles.

Key to species. Kahl (1930–35).

Descriptions of species. Penard (1922), Prelle (1968), Wenzel (1953).

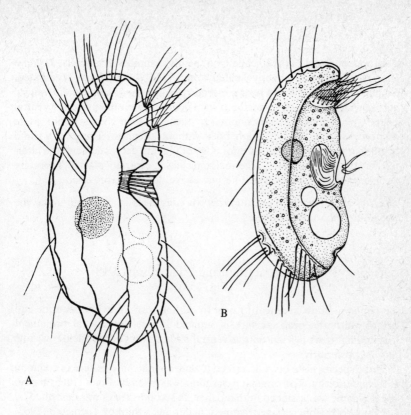

Fig. 161. *Drepanomonas.* A, After Wenzel (1953). B, After Mermod (1914).

Hemicyclium Eberhard, 1862
(Fig. 162)

Description. Body outline shape oval, highly compressed laterally, covered with rigid, delicate, slightly keeled armour. Ribs on armour crenulated. Somatic ciliature reduced to three semicircular rows of cilia on the ventral surface. Opening to the cytostome via a depression on the posterior ventral edge. The depression opens more to the side than the posterior as in *Microthorax* (p. 260). Cytopharynx without trichites. There is a solid tooth-like structure on the anterior edge of the oral depression and the latter is ciliated on its right side. The macronucleus is centrally placed. There are two contractile vacuoles which lie just above the oral depression.

Description of species. Penard (1922) who describes an organism under the name *Microthorax haliotideus*.

Hexotricha Conn and Edmondson, 1918
(Fig. 163)

Description. Body spherical to oval in outline with lip at the anterior end below which the oral aperture is situated. There are several prominent longitudinal rows of cilia on the ventral surface. Single contractile vacuole located posteriorly.

Three species have been described to date and the only feature in common is the anterior lip. The original description was given by Conn (1905) but no generic name was assigned in that paper. A second species was described by Lackey (1925) from sewage-treatment plants and a third by Tucolesco (1962) from cave pools.

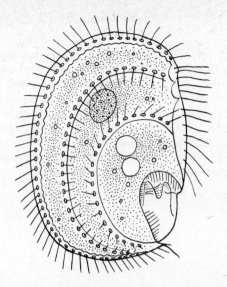

Fig. 162. *Hemicyclium* (after Penard, 1922).

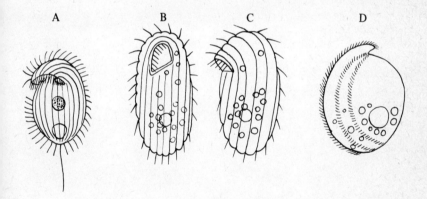

Fig. 163. *Hexotricha*. A, After Lackey (1925). B, Ventral aspect. C, Lateral aspect (both after Tucolesco, 1962). D, after Conn (1905).

Kreyella Kahl, 1931
(Fig. 164)

Description. Body outline shape oval, flattened with delicate keeled armour. Somatic ciliature reduced, ventral surface with two or three semicircular ciliary rows. Opening to cytostome in a depression situated in the posterior left region of body. Oral aperture not supported by basket of trichites. On the right of the oral depression there is a posterior vibratile edge and there are two parallel anterior ciliary rows. On the left of the oral aperture there is a line of membranelles arising from paired rows of kinetosomes. Macronucleus rounded and centrally located. Single contractile vacuole on posterior right of body.

Descriptions of species. Foissner (1979), Kahl (1931).

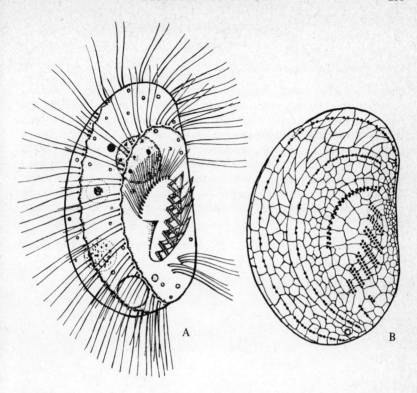

Fig. 164. *Kreyella*. A, Whole animal. B, Silver-impregnated specimen (after Foissner, 1979).

Leptopharynx Mermod, 1914
(Fig. 165)

Trichopelma Levander, 1900

Description. Body shape irregularly oval in outline, laterally compressed and covered with a longitudinally ridged rigid pellicle. Somatic cilia not dense but they cover the entire body. There are four kineties on each side of the body but are rather more regularly arranged on the right side. Cytostome close to body edge, cytopharynx tubular supported by a basket of trichites. On the right and above the cytostome is a short row of double cilia (kinety one) forming the paroral ciliature and below the cytostome are two membranelles. Trichocysts and/or zoochlorellae present in some species. Single contractile vacuole. Spherical macronucleus centrally situated with an adjacent micronucleus.

Descriptions of species. Kahl (1930–35), Mermod (1914), Njiné (1980), Prelle (1961), Savoie (1957).

Taxonomy. Prelle (1962).

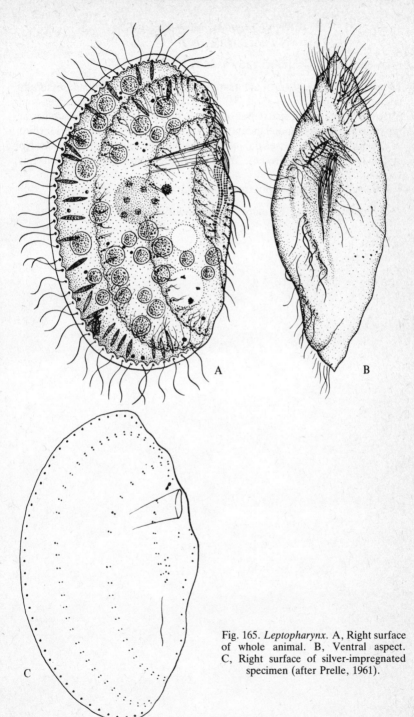

Fig. 165. *Leptopharynx*. A, Right surface of whole animal. B, Ventral aspect. C, Right surface of silver-impregnated specimen (after Prelle, 1961).

Microdiaphanosoma Wenzel, 1953
(Fig. 166)

Diaphanosoma Grandori and Grandori, 1934

Description. Body shape oval (rarely sickle shape) in outline. Laterally compressed, encased in rigid pellicle which has longitudinal ridges and furrows. Somatic cilia sparse but uniformly distributed. Oral aperture lies at the base of a deep rounded indentation which is situated in the anterior third of the body. Both edges of the indentation are lined with a row of cilia. No cytopharyngeal basket present. There is a single contractile vacuole which is terminal.

Rare, described once only from soils in Italy.

Description of species. Grandori and Grandori (1934).

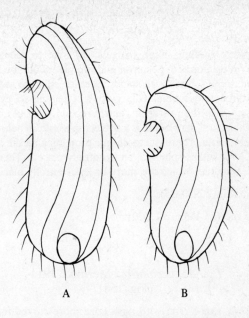

A   B

Fig. 166. *Microdiaphanosoma*. A, B, Variation in size and shape (after Grandori and Grandori, 1934).

Microthorax Engelmann, 1862
(Fig. 167)

Description. Body outline shape irregular to oval to almost triangular, anterior more or less pointed, posterior end rounded. Flattened with delicate keeled armour, some species with spines. Somatic ciliature reduced, ventral surface with three longitudinal ciliary rows. Opening to cytostome in a depression situated in the posterior ventral region, without cytopharyngeal basket. On right of oral depression is a stiff ectoplasmic lip below which there is a small membranelle. There is a tooth-like projection on the left side of the cytostome. Macronucleus spherical and centrally placed. There are usually two contractile vacuoles which are more or less centrally placed.

Key to species. Kahl (1930–35).

Description of species. Penard (1922).

Pseudomicrothorax Mermod, 1914
(Fig. 168)

Description. Body shape oval with right edge more curved than left edge, dorso-ventrally flattened. Covered with rigid and inflexible pellicle. Somatic ciliature in form of several (approx. 15) longitudinal ciliary meridians, more meridians on ventral than dorsal surface. Ventral cilia longer and finer than dorsal cilia. Oral aperture in anterior third of body lying within very slight depression. Cytopharynx supported by basket of delicate trichites. There are three compact groups of membranelles on the left of the cytostome and a short curved row of long cilia on its right. The latter, which partly encircle the cytostome, do not coalesce to form an undulating membrane but remain free. Beneath this row is a group of four cilia and there is another group of four close to the opening of the canal which leads to the centrally placed contractile vacuole. There are many trichocysts below the pellicle. Macronucleus elongate ovoid, centrally placed with single micronucleus.

Descriptions of species. Mermod (1914), Peck (1975), Thompson and Corliss (1958).

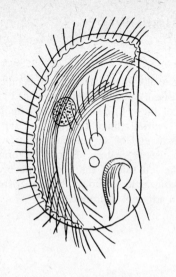

Fig. 167. *Microthorax* (composite after Kahl, 1930–35 and Bick, 1972).

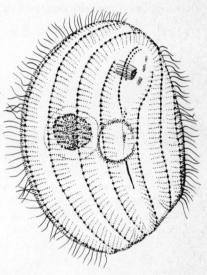

Fig. 168. *Pseudomicrothorax* (composite from Corliss, 1979 and Jankowski, 1964).

Stammeridium Wenzel, 1969
(Fig. 169)

Stammeriella Wenzel, 1953

Description. Body outline shape elongate oval with notches and projections at each end. Rigid body, laterally compressed. Somatic ciliature reduced to few (about 12) long cilia at either end of the cell, cilia borne only upon right-hand surface except for the single ventral cilium. Cytopharynx supported by trichites, oral aperture opens at end of an anteriorly situated ventral projection which is the major generic feature. Below the oral aperture there is a line of fine cilia. The left side of the body is deeply furrowed longitudinally. There are two contractile vacuoles. The spherical macronucleus is centrally situated with an adjacent micronucleus.

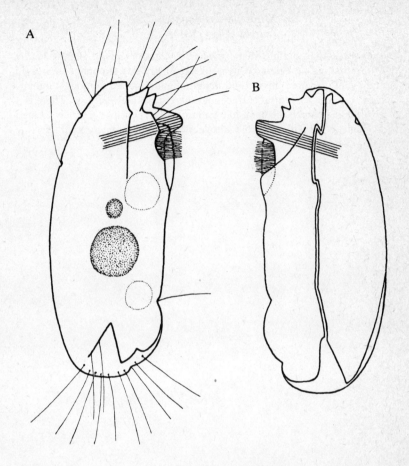

Fig. 169. *Stammeridium*. A, Right lateral surface. B, Left lateral surface (after Wenzel, 1953).

Trochiliopsis Penard, 1922
(Fig. 170)

Description. Body outline shape oval with anterior curved slightly to the right terminating in a pointed beak-like region. Body strongly compressed laterally and covered in a rigid longitudinally ribbed armour. Somatic ciliation reduced to ventral surface and these arise between ribs. There is a group of cilia beneath the beak-like anterior. Macronucleus centrally located with single contractile vacuole above it. There are many characteristic trichocysts present.

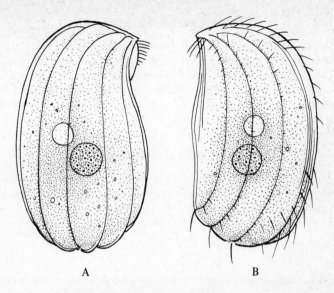

A   B

Fig. 170. *Trochiliopsis*. A, Dorsal surface. B, Ventral surface (after Penard, 1922).

Order CYRTOPHORIDA

The hypostomial frange is not identifiable as such in this order but the oral area contains several short double rows of kinetosomes located anteriorly to a complicated cyrtos with the pre-oral suture being skewed far to the left. The body is often dorsoventrally flattened with the somatic ciliation being restricted to the ventral surface. In some there is an adhesive organelle in the form of a posterior protoplasmic spine.

Suborder CHLAMYDODONTINA

These genera have the characters of the above order. The ventral ciliation is thigmotactic but there is no posterior protoplasmic spine for adhesion. The body is generally broad in shape and dorso-ventrally flattened.

Chilodonatella Dragesco, 1966
(Fig. 171)

Description. Body outline oval to pyriform, without obvious dorso-ventral flattening. Rather small (20 μm long). Somatic ciliature restricted to the ventral surface which is longitudinally ridged. The ciliature is highly simplified and consists of four parallel rows of cilia on either side of the body which curve around to meet anteriorly in front of the oral aperture. The aperture is rounded and is supported by a basket of trichites. The macronucleus is large, ventrally located in the posterior body third. There is a single large contractile vacuole which lies dorsally in the posterior body third.

Description of species. Dragesco (1966a).

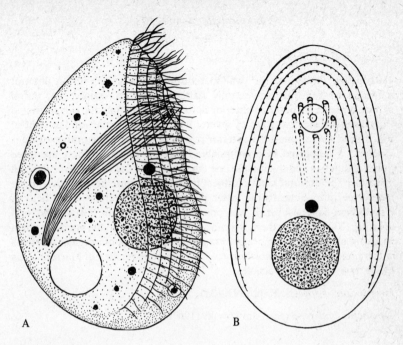

Fig. 171. *Chilodonatella*. A, Lateral surface. B, Ventral surface (after Dragesco, 1966a).

Chilodonella Strand, 1928
(Fig. 172)

Chilodon Ehrenberg, 1838

Description. Outline shape oval to reniform sometimes with an anterior left-hand rostrum. Dorso-ventrally flattened, ventral surface flat, dorsal surface arched except for an anterior flattened region. Somatic cilia restricted to ventral surface consisting of several longitudinal kineties on the right of which most curve round the anterior region of the cell. The oral aperture is oval and is supported by a protrusible basket of trichites. There are three pre-oral kincties and the middle one curves around the right side of the oral aperture. The pre-oral kinety furthest from the aperture lies obliquely along the suture line. Beneath the latter pre-oral kinety there is a left field of longitudinal kineties which are positioned only on the left of the oral aperture. That is to say there is a central zone posterior to the aperture which is devoid of cilia, silver impregnation reveals a network of irregular polygons in this region. In *Trithigmostoma* (p. 280) this zone is ciliated. Macronucleus ovoid, two contractile vacuoles.

Description of species. Kahl (1930–35).

Taxonomy and synonymy. Jankowski (1967b).

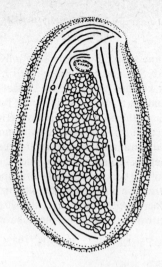

Fig. 172. *Chilodonella*. Ventral surface of a silver-impregnated specimen (after Jankowski, 1967b).

Gastronauta Bütschli, 1889
(Fig. 173)

Description. Body outline shape oval, dorso-ventrally flattened with dorsal surface strongly arched. Somatic ciliature reduced to ventral surface with exception of two short kineties on the dorsal surface. The oral aperture is very long and narrow and lies transversely across the body, it is not supported by trichites. On the right of the body there are several longitudinal kineties of which three or four extend forward and curve around the apex. Between these curved ends and the oral aperture are several curved pre-oral kineties. There is a pre-oral kinety which entirely surrounds the oral aperture, this is the diagnostic feature of the genus. Below the oral aperture there is the left field of kineties which traverse the body width (although there may be a large gap in the mid-line). Macronucleus large, irregular shape located in posterior half of body. There are two contractile vacuoles, one anterior, one posterior.

Descriptions of species. Pätsch (1974), Wilbert (1972).

Taxonomy. Deroux (1970, 1976).

Fig. 173. *Gastronauta*. A, Lateral view (from life). B, Dorsal aspect showing some ventral ciliation. C, Ventral aspect of silver-impregnated specimen (partly from life, partly from Wilbert, 1972).

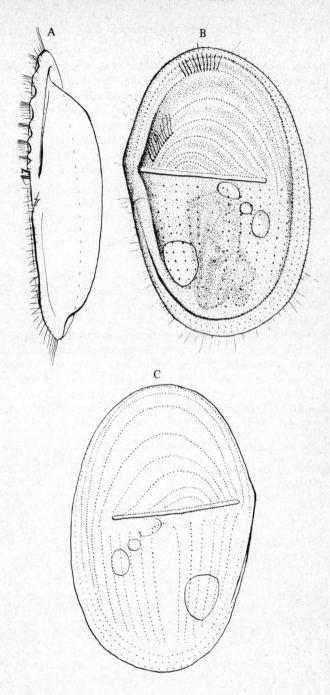

Lophophorina Penard, 1922
(Fig. 174)

Description. Body outline oval to reniform, dorso-ventrally flattened with strongly arched dorsal surface. Somatic ciliature restricted to ventral surface where cilia arise from several curving longitudinal striations. Oral aperture in anterior region but without a basket of trichites. A flabby mobile tentacle-like structure composed of several cilia bound together arises from the left side of the apex. Macronucleus elongate and horseshoe shape. Two contractile vacuoles present, one anterior, one posterior. Description based on single report by Penard (1922) who observed the animal as a commensal on the crustacean *Gammarus* in Switzerland.

Chlamydodon Ehrenberg, 1835
(Fig. 175)

Description. Outline shape oval to reniform, dorso-ventrally flattened with edges of the dorsal body half overhanging the ventral half. Between the two halves is a prominent band of trichites (sometimes referred to as the 'railroad track') which encircles the body. Somatic ciliature restricted to ventral surface. Ventral ciliature consists of many longitudinal kineties on the right-hand side which extend forward to curve around the apical region. The oral aperture is supported by a ring of trichites. There are three pre-oral kineties. Behind the mouth there is a field of left kineties which completely fills the left half of the ventral surface. The macronucleus is ovoid and centrally located. There are several contractile vacuoles scattered throughout the cytoplasm.

Keys to species. Kahl (1930–35), Kaneda (1953).

Descriptions of species. Fauré-Fremiet (1950), Kahl (1930–35), Kaneda (1953, 1960).

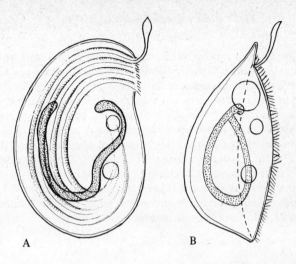

Fig. 174. *Lophophorina*. A. Ventral view. B, Lateral view (after Penard, 1922).

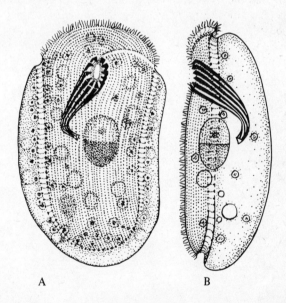

Fig. 175. *Chlamydodon*. A, Ventral aspect. B, Lateral aspect (after Kaneda, 1953).

Chlamydonella Deroux, 1970
(Fig. 176)

Description. Oval in outline shape, dorso-ventrally flattened with dorsal surface strongly arched. Somatic ciliation reduced to ventral surface with a short field of cilia just on the left of the anterior dorsal surface. Oral aperture supported by basket of trichites. On the right of the ventral surface there are several kineties of which two curve around the apex. The extreme right kinety is discontinuous in two parts, one in the equator the other just dorsally on the extreme left of the apex. Below the oral aperture is the left field of kineties (10 or more) which traverse the body width. The arrangement of the pre-oral ciliature is a major characteristic of the genus, the kinety just in front of the oral aperture is always Y-shaped and above it lies a second pre-oral kinety which is simply curved. Macronucleus large, centrally located. There are two contractile vacuoles, one anterior, one posterior.

Descriptions of species. Foissner (1980), Pätsch (1974).

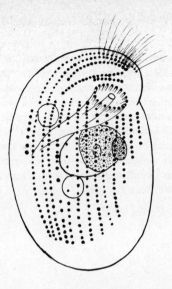

Fig. 176. *Chlamydonella*. Ventral surface of silver-impregnated specimen (after Patsch, 1974).

Odontochlamys Certes, 1891
(Fig. 177)

Description. Outline shape oval to pyriform, convex on left side, concave on right. Dorso-ventrally flattened, ventral surface flat, dorsal surface strongly arched. Genus easy to identify because of the bump-like spines (four to thirteen) which project out from the dorsal surface around the periphery of the cell. Somatic ciliature restricted to ventral surface where there is a simplification as in *Chilodonatella* (p. 266). There are four symmetrical longitudinal rows of cilia on either side of the body which curve round to meet anteriorly. The rounded oral aperture is supported by a basket of trichites. Macronucleus ovoid, centrally placed. There are three contractile vacuoles scattered throughout the cell.

Description of species. Certes (1891).

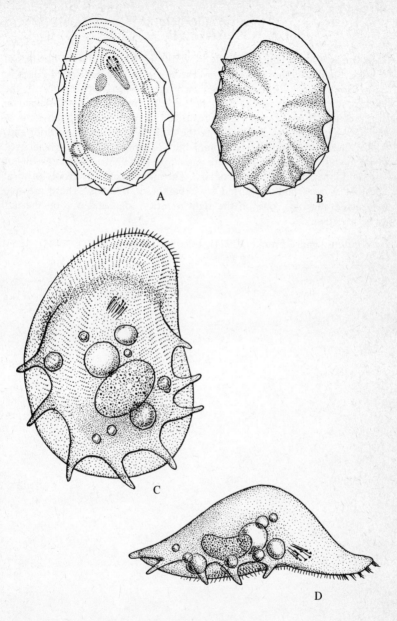

Fig. 177. *Odontochlamys*. A, Ventral surface of silver-impregnated specimen. B, Dorsal surface (both after Wenzel, 1953). C, Dorsal surface. D, View of right side (both after Certes, 1891).

Phascolodon Stein, 1859
(Fig. 178)

Description. Outline shape of body viewed from dorsal surface is like that of a tulip. Slightly compressed dorso-ventrally with well-developed edges of body forming wing-like extensions which turn towards each other over the ventral surface. This means that the body is approximately C-shaped in cross-section. Somatic cilia mainly restricted to ventral surface but some of the right hand kineties do cross onto the dorsal surface at the anterior end. These then curve completely around the C-shaped apical edge and back down the left ventral surface. There appears to be a central area between left and right kineties that is devoid of cilia. The macronucleus is ovoid. The oral aperture is oval and supported by a basket of trichites. There are two contractile vacuoles, one on the right anterior edge and one on the left posterior edge.

Descriptions. Fauré-Fremiet (1924), Foissner (1980), Kahl (1930–35), Stein (1859a).

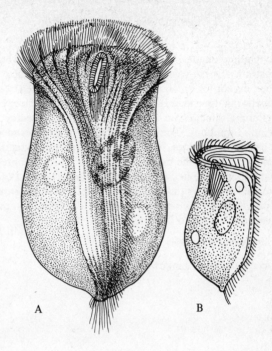

Fig. 178. *Phascolodon*. A, Ventral view (after Fauré-Fremiet, 1924). B, Side view (after Kahl, 1930–35).

Trithigmostoma Jankowski, 1967
(Fig. 179)

Description. Outline shape oval to reniform, sometimes with an anterior left-hand rostrum. Dorso-ventrally flattened, ventral surface flat, dorsal surface arched except for an anterior flattened region which bears a row of bristles. Somatic ciliature mainly restricted to ventral surface, consisting of several longitudinal kineties on the right of which most curve around the anterior region of the body. The oral aperture is oval and is supported by a protrusible basket of trichites. There are three pre-oral kineties, and the middle one curves around the right side of the aperture. The pre-oral kinety furthest from the aperture lies along the suture line. The rest of the ventral surface is covered with longitudinal kineties and there is no central cilia-free zone as there is in *Chilodonella* (p. 268). Macronucleus ovoid, many contractile vacuoles scattered throughout the cell.

Key to species. Kahl (1930–35).

Taxonomy and synonymy. Jankowski (1967b).

Fig. 179. *Trithigmostoma*. A, Ventral aspect. B, Side aspect (after Mackinnon and Hawes, 1961). C, Silver-impregnated ventral surface (after Jankowski, 1967b).

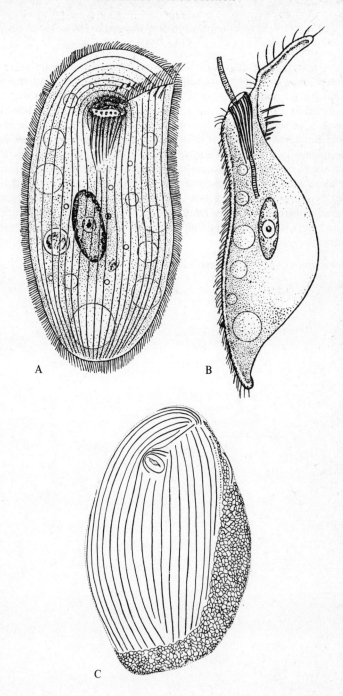

A

B

C

Pseudochilodonopsis Foissner, 1979
(Fig. 180)

Description. Outline shape oval to reniform, often with an anterior left-hand rostrum. Dorso-ventrally flattened, ventral surface flat, dorsal surface arched except for an anterior flattened region. The oral aperture is oval and is supported by a basket of trichites. Somatic cilia restricted to ventral surface consisting of several longitudinal kineties on the right of which most curve round the anterior region of the cell. On the left there is another field of longitudinal kineties which are separated from the right field by a large post-oral gap. There are two pre-oral kineties close to the oral aperture and the more anterior one curves around it. There are also four other short kineties lying along the suture line which is the genetic characteristic separating it from the closely related genus *Chilodonella* (p. 268) which has a single kinety stretching along the suture line. Macronucleus ovoid; with two contractile vacuoles.

Description of species. Foissner (1979).

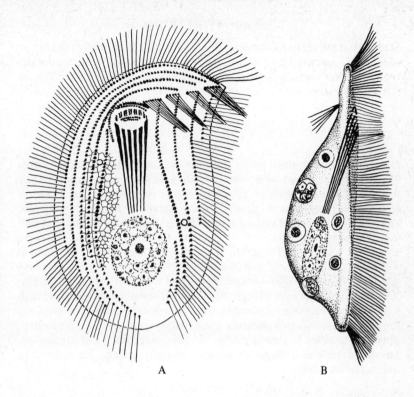

A   B

Fig. 180. *Pseudochilodonopsis*. A, Ventral view. B, Side view (after Foissner, 1979).

Suborder DYSTERIINA

Members of this suborder have the characteristics of the order but the basket of trichites supporting the oral area tends to be specialised. In the freshwater representatives there is a posterior protoplasmic spine which secretes an adhesive thread.

Dysteria Huxley, 1857
(Fig. 181)

Description. Body outline shape approximately oval with the dorsal surface being strongly arched and often longitudinally ribbed, ventral surface flattened. There is a posterior cytoplasmic spine from which a thread may be secreted for attachment to the substratum. Somatic ciliation greatly reduced and restricted to ventral surface alone. The right ciliary field consists of four to six kineties of which two to four extend beyond the oral aperture and curve around the body apex. The left field of kineties is in two parts as is found in *Trochilia* (p. 288) and not as a continuous field as in *Orthotrochilia* (below) and *Trochilioides* (p. 288). One part consists of a few rows lying equatorially, close to and parallel with the right field. The other part of the left field lies in front of the oral aperture situated between the pre-oral and circumoral ciliary rows. There are only two elements in the cytopharyngeal basket, the teeth of which are complex in structure (Fig. 181) but never sharp and needle-like. The macronucleus is large, ovoid and centrally placed. There are two contractile vacuoles.

Key to species. Kahl (1930–35).

Descriptions of species. Deroux (1965), Fauré-Fremiet (1965).

Taxonomic review. Deroux (1977).

Orthotrochilia Deroux, 1977
(Fig. 182)

Description. Body outline shape oval, dorsal surface strongly arched, ventral surface flattened. There is a posterior cytoplasmic spine on the ventral surface from which a secreted thread may attach the animal to the substratum. Somatic ciliature severely reduced and located only on the ventral surface. As in *Trochilioides* (p. 288) there are fields of both right and left cilia. Those on the left are short and extend across the ventral surface beneath the oral apparatus. The elements of the oral basket are not reduced as in *Trochilioides* but consist of numerous stylets without teeth (Fig. 182). Macronucleus large, ovoid and of the heteromeric type. There are two contractile vacuoles.

This is mainly a marine genus but one freshwater species has been noted – *Orthotrochilia* (*Trochilioides*) sp. Pätsch, 1974.

Taxonomy. Deroux (1977).

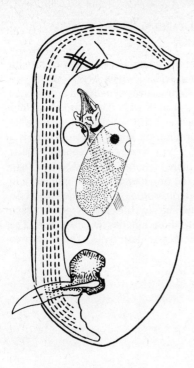

Fig. 181. *Dysteria*. Ventral surface, Silver-impregnated specimen (after Fauré-Fremiet, 1965).

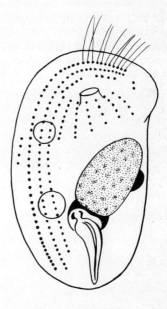

Fig. 182. *Orthotrochilia*. Silver-impregnated specimen, ventral surface (after Pätsch, 1974).

Parachilodonella Dragesco, 1966
(Fig. 183)

Description. Body shape oval to reniform, dorso-ventrally flattened with dorsal surface more arched than ventral surface which bears one or two posterior cytoplasmic spines. Somatic ciliature severely reduced and located only on ventral surface except for a few dorsal bristles. There are about four kineties on the right-hand side, of which at least two travel the complete length of the body curving anteriorly around the apical region. There is a separate left-hand field of kineties which may traverse the complete body width but in one species (a marine one) there is a large gap between the right and left fields of cilia. The elements of the oral basket are not reduced and there are many trichites as is found in *Orthotrochilia* (p. 284). The transverse kineties in front of the oral aperture lie side by side in this genus which distinguishes it from *Orthotrochilia* where those kineties lie end to end. The macronucleus is large, ovoid and of the heteromeric type. There are two contractile vacuoles.

Description of species. Wilbert (1972).

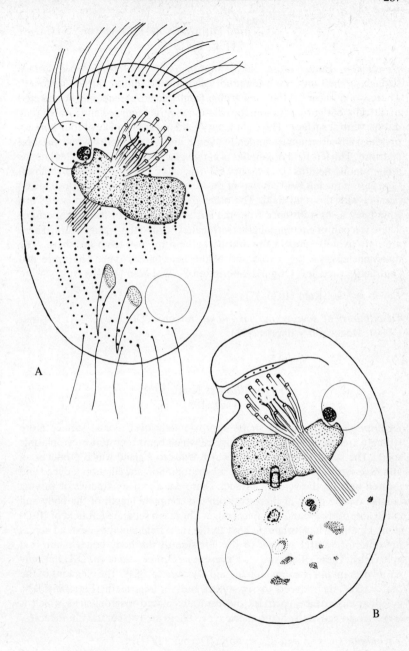

Fig. 183. *Parachilodonella*. A, Ventral surface, silver impregnation. B, Dorsal surface (after Wilbert, 1971).

Trochilia Dujardin, 1841
(Fig. 184)

Description. Body outline shape approximately oval with dorsal surface strongly arched and often longitudinally ribbed, ventral surface flattened. There is a posterior cytoplasmic spine from which a thread may be secreted for attachment purposes. Somatic ciliation severely reduced and located only on the ventral surface. The right ciliary field always consists of four kineties of which three terminate under the spine and two extend beyond the oral aperture. The left field of kineties is in two parts as in *Dysteria* (p. 284) not as a continuous field as in *Orthotrochilia* (p. 284) and *Trochilioides* (below). One part of the left field consists of three rows lying equatorially, close to and parallel with the right field. The other part of the left field consists of two rows lying a short distance from the left anterior edge of the oral aperture. There is a pair of circumoral kineties located beyond the oral aperture. There are only two elements in the oral basket and the two teeth are sharp. The macronucleus is large, ovoid and of the heteromeric type. There are two contractile vacuoles. Original description by Dujardin (1841b).

Key to species. Kahl (1930–35).

Descriptions of species and taxonomic revision. Deroux (1977), Foissner (1980), Heuss and Wilbert (1973).

Trochilioides Kahl, 1931
(Fig. 185)

Description. Body outline shape approximately oval, dorsal surface more strongly arched than the ventral surface which bears a posterior cytoplasmic spine. The spine contains a canal which leads to a gland which produces an adhesive secretion for fixation to a substratum. Somatic ciliation reduced and located only on the ventral surface. There are about six kineties of varying lengths on the right; of these two run the complete length of the body and curve anteriorly around the apical edge. There is a separate left field of 10–20 short kineties which originate next to the right field and these extend across the anterior ventral surface to the left side of the body beneath the oral apparatus. These ciliary rows are present in *Orthotrochilia* (p. 284) but only partly present in *Trochilia* (above) and *Dysteria* (p. 284). The elements of the oral basket are reduced to six stylets and six long teeth (Fig. 185). The cytopharyngeal tube encircles the centrally placed macronucleus which is large, ovoid and of the heteromeric type. There are two contractile vacuoles.

Taxonomic revision and descriptions. Deroux (1977).

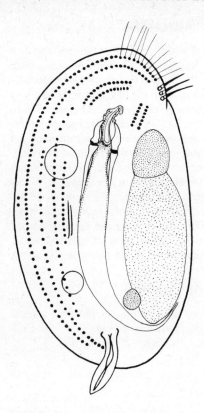

Fig. 184. *Trochilia*. Silver-impregnated specimen, ventral surface (after Heuss and Wilbert, 1973).

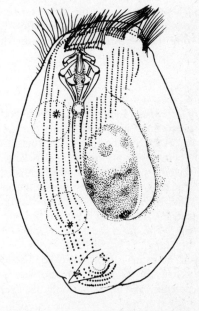

Fig. 185. *Trochilioides*. Ventral surface, silver-impregnated specimen (after Deroux, 1976c).

Order CHONOTRICHIDA

The chonotrichs form a small group of mainly marine organisms although there are three freshwater genera. Of the three, only *Spirochona* has been reported in Europe, the others seem to be limited to Lake Baikal in Russia. All chonotrichs live epizooically on the mouthparts of crustaceans to which they may be attached directly or by a non-contractile stalk. There is no stalk in the freshwater representatives. At the opposite end of the base-like body there is an expanded funnel region (pre-oral funnel) which represents an expanded vestibulum (commonly referred to as a peristome) lined with cilia. While these cilia are extensions of the somatic ciliature, the latter is not present in the adult. In all the freshwater genera the pre-oral funnel is a distinctly spiral structure. Chonotrichs may reproduce sexually by conjugation (Tuffrau, 1953) but more commonly they do so asexually by budding. Budding may be internal or external giving rise in both cases to ciliated larval forms which are remarkably like those of suctoria (Guilcher, 1951). Exogenous budding is the only type represented in the freshwater genera.

Suborder EXOGEMMINA

These genera have the characteristics of the above order but with asexual budding limited to the exogenous type producing a single bud. Attached directly to crustacean mouthparts without a stalk.

Cavichona Jankowski, 1973
(Fig. 186)

Description. Elongate, cylindrical body, rounded in cross-section attached to crustaceans by a distinct and elongate narrow region of the body (pseudo-style) which is normally free from food vacuoles. There are no somatic cilia. The anterior pre-oral funnel is large, ciliated on the inner surface, with a distinct spiral-like portion which is composed of several (up to five) whorls. The funnel is the widest part of the animal and there is a narrow neck region. The ventral wall (rim of outer whorl) of the pre-oral funnel is interupted by a single notch so that there are two discontinuous ledge-like folds on the whorl rim. The macronucleus is rounded and reproduction is by exogenous budding.

The genus is most easily confused with *Spirochona* (p. 292) which has no notches in the whorl rim and with *Serpentichona* (p. 292) which has many notches. The genus has only been recorded in Lake Baikal.

Key to species. Jankowski (1973b).

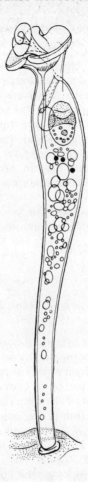

Fig. 186. *Cavichona* (after Jankowski, 1973b).

Serpentichona Jankowski, 1973
(Fig. 187)

Description. Very elongate, cylindrical body attached by aboral end to crustaceans. The posterior region (pseudostyle) is distinctly elongate, narrow and free from any food vacuoles. There are no somatic cilia. The pre-oral funnel is large, wide but comparatively low and there is a tendency towards the underdevelopment of the whorl region (usually less than two and a half whorls but up to three and a half whorls). The rim of the pre-oral funnel is broken by several notches so there are several (up to eight) discontinuous ledge-like folds on the whorl rim. In addition the body has several longitudinal rib-like folds. The macronucleus is rounded and reproduction is by exogenous budding.

The genus is most easily confused with *Spirochona* (below) and *Cavichona* (p. 290). *Serpentichona* has only been recorded in Lake Baikal.

Key to species. Jankowski (1973b).

Spirochona Stein, 1852
(Fig. 188)

Description. Elongate, rather cylindrical body attached by aboral end to crustacea. There is an anterior pre-oral funnel which is ciliated on the inner surface, with a spiral-like portion which has few (two to three) whorls. The funnel is generally the widest part of the body and it narrows to a short neck region before joining the rest of the cell. There are no somatic cilia. The ventral wall of the pre-oral (rim of the outer whorl) is without folds or notches so that the margin is smooth and there is a single continuous ledge-like rim to the whorl. These features serve to distinguish the genus from *Cavichona* (p. 290) and *Serpentichona* (above). The macronucleus is rounded and reproduction is by exogenous budding.

Keys to species. Jankowski (1973b), Swarczewsky (1928e).

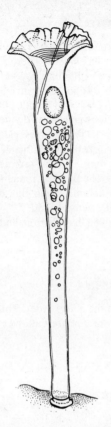

Fig. 187. *Serpentichona* (after Jankowski, 1973b).

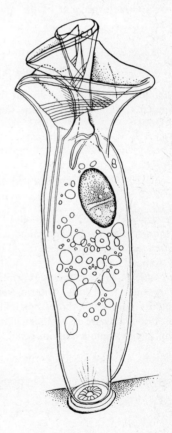

Fig. 188. *Spirochona* (after Jankowski, 1973b).

Order APOSTOMATIDA

The apostomes typically comprise a marine group, many being truly parasitic, but all are closely associated with invertebrate animals. There are three freshwater genera which while closely associated with crustaceans are never truly parasitic. The feeding stage or trophont feeds upon the exuvial fluid of the moulted exoskeleton of its 'host'; food being ingested via the characteristic rosette-like cytostome. The trophont grows but cannot divide. After feeding the cell encysts and undergoes division to produce several tomites which are free-swimming stages which serve to distribute the organism from 'host' to 'host'. When a new 'host' is found, the tomite settles, secretes a cyst (often on a short peduncle) to form a phoront. The phoront remains in the encysted state until the 'host' moults its exoskelton. When this happens trophonts emerge from the cyst to feed upon the exuvial fluid.

A complete account of the apostomes is given by Chatton and Lwoff (1935).

Suborder APOSTOMATINA

Members of this suborder display the features of the above order. This group is mainly marine but contains the only freshwater representatives of the order.

Gymnodinioides Minkiewicz, 1912
(Fig. 189)

Description. Trophont. In moult of freshwater gammarids *Gammarus pulex*, *Echinogammarus berillono* and *Carinogammarus roeseli* and on *Asellus aquaticus*. Body twisted but ovoid to pyriform in shape. There are nine rows of somatic cilia which spiral dextrally from anterior to posterior. Eight originate at or near the apex while kinety one originates about quarter of body length from apex. Anterior ventral surface without clumps of cilia. Oral ciliature consists of three short ciliary rows (xyz) left and posterior to the rosette. There are two other ciliary rows (α and β) anterior to rows xyz. Contractile vacuole without canal. Macronucleus elongate.

Phoront. Cyst attached to crustaceans on branchial lamellae of pleopods.

Description and life cycle. Chatton and Lwoff (1935).

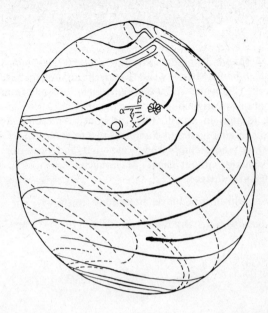

Fig. 189. *Gymnodinioides* showing spiral ciliary meridians (after Chatton and Lwoff, 1935).

Hyalophysa Bradbury, 1966
(Fig. 190)

Description. Trophont. In moult of freshwater shrimp *Palaemonetes paludosus* and crayfish *Cambarrus*. Body twisted but pyriform in shape. There are nine rows of somatic cilia which spiral dextrally down the body from anterior to posterior. Eight rows originate at or near the apex but kinety nine originates near rosette located in middle third of body. Anterior ventral surface flattened and contains scattered clumps of two or three cilia. Oral ciliature consists of three short ciliary rows (xyz) left and posterior to rosette. Contractile vacuole without canal. Macronucleus elongate, situated just beneath midventral surface.

Phoront. Cyst with short peduncle attached to crustacea.

Description and life cycle. Bradbury and Clamp (1973), Grimes (1976).

Hyalospira Miyashita, 1933
(Fig. 191)

Description. Trophont. In moult of freshwater shrimp *Xiphocaridina compressa* but also seen on other freshwater crustaceans such as *Leander paudicens, Macrobrachium nipponense* and even on moult of a mayfly nymph. Body twisted but usually reniform in shape. There are nine rows of somatic cilia which spiral dextrally down the body from anterior to posterior. Rows one and nine do not originate at the apex but the other seven do. (NB original author numbered somatic rows incorrectly.) The postoral ciliation consists of two groups; there are three short rows (xyz) behind the rosette and two on its right. There is a single large, centrally placed contractile vacuole with a very long anteriorly directed accessory canal. The single elongate macronucleus lies close to the rosette.

Phoront. Cysts found attached to the squamae and basal part of the second antennae and the branchial lamellae of freshwater crustaceans.

Description and life cycle. Miyashita (1933).

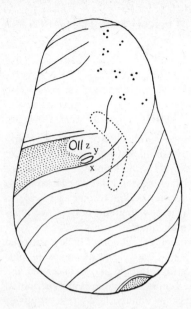

Fig. 190. *Hyalophysa* showing spiral ciliary meridians (after Grimes, 1976).

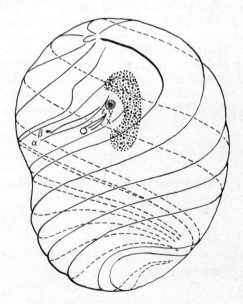

Fig. 191. *Hyalospira* showing spiral ciliary meridians (after Miyashita, 1933).

Keys and descriptions of genera of Suctoria

Subclass SUCTORIA

At first sight the adult suctorian would appear to have little in common with the rest of the ciliated protozoa. Closer examination reveals however, that while the adults are without cilia they do retain the kinetosomes which are scattered throughout the cytoplasm and then passed on to the larval forms during budding. Furthermore in these larvae the kinetosomes give rise to cilia in the normal way. In addition both larvae and adults possess two types of nucleus which is a feature common to all ciliates.

The most obvious suctorian feature is their possession of **hollow suctorial tentacles.** These may have knobs on their ends like the heads of pins (capitate), may be pointed, may bend or be retractile like a concertina and rarely may have distensible tips. In some genera the tentacles are distributed all over the body surface while in others they emerge in discrete groups or bundles known as **fascicles.** While tentacles vary in their appearance and mode of action they are all concerned with food capture and ingestion. The majority of suctoria feed on other organisms, particularly other protozoa which are caught and narcotised by the tentacles and finally the cell contents of the prey are sucked down through the hollow tentacles into the body of the suctorian. In addition to this mode of life some suctoria are parasitic and are therefore not described here, however, the freshwater parasitic genera include *Endosphaera* Engelmann, 1876; *Kystopus* Jankowski, 1967a; *Pottsiocles* Corliss, 1960 and *Pseudogemmides* Kormos, 1938.

The majority of suctoria attach themselves, by some means, to a variety of submerged objects including a whole range of animal and plant life. Attachment may be by means of a **stalk** which term here is limited to describing those elongate attachment organelles which are completely separated off from the body and are not extensions of the pellicle or of the lorica. In other suctoria, the body may attach directly to the substratum without the presence of an intervening stalk or stalk-like extension of the lorica. Many suctoria live in membranous shell-like protective secretions known as loricas. There is a good deal of confusion in the literature as to what constitutes a lorica. In the French literature two terms 'loge' and 'coque' have been used (Collin, 1912), the former type is usually closely fitting and there is never a lateral free margin between the body and the lorica. The latter term 'coque' is used to describe those loricas that are much more like those of the peritrichs where there is a free lateral margin between body and lorica. Finally, other suctoria have pellicles which unfortunately tend to merge with the 'loge' type loricas and are sometimes difficult to distinguish.

The simple morphology of the adult was the first taxonomic feature to be used for the identification and classification of suctoria. Later it was found that the ciliated larvae were formed by two major processes, either by internal (endogenous) budding or by external (exogenous) budding. Since then variations on these two major types have been described (Batisse, 1975a); some suctoria produce more than one bud at a time (multiple budding) and in some cases a bud, produced internally with nuclei and cilia, may not separate from the mother cell until after it has been evaginated. All of these types of asexual reproductive processes are now included in taxonomic studies and are used as the basis of classification (see Batisse, 1975a,b).

The morphology of the larva is also now widely used (Canella, 1957; Kormos and Kormos, 1956a,b,c, 1957, 1958a,b, 1960a,b, 1961; Matthes, 1956) for taxonomic purposes although in many cases these features are used more for familial characters rather than for generic ones.

In the key which follows an attempt has been made to use the morphological features of the adult as far as is possible. In most cases these features are still sufficient to identify a genus as it was originally defined, however, in a few cases the type of budding needs to be established, or a knowledge of the type of larva is necessary. In any case it should be emphasised that wherever possible identifications should be confirmed by a study of the animal's life cycle. The important work by Matthes (1954a,b,c,d,e) has clearly demonstrated the variability of the morphology of suctoria during the life cycle of *Discophrya* and these studies throw serious doubts on the validity of classical suctorian taxonomy based on morphological features alone. More work of that nature is required, it is hoped that this key will stimulate more workers to enter the difficult field of suctorian taxonomy which needs extensive revision.

Key to the genera of Suctoria

1. Attached to substratum **5**
   Not attached to a substratum **2**

2. Body with six or eight arm-like extensions which bear tentacles .. **4**
   Body without arm-like extensions **3**

3. Body spherical, without mucus covering .... *Sphaerophrya* (p. 316)
   Body shape irregular, with thick transparent mucus coat
   .... *Mucophrya* (p. 312)

4. Body with six arm-like extensions, not covered in sand grains
   .... *Staurophrya* (p. 338)
   Body with eight arm-like extensions, may be covered in sand
   grains .... *Astrophrya* (p. 322)

5. Attached to substratum by means of one or two posterior protoplasmic processes **6**
   Attached to substratum by other means **7**

6. Attached to peritrich colony by means of two posterior protoplasmic processes .... *Erastophrya* (p. 330)
   Attached to inanimate substratum by means of a single protoplasmic process .... *Brachyosoma* (p. 324)

7. Attached by means of a stalk or stalk-like extension of lorica **8**
   Body or lorica attached directly to substratum without an intervening stalk or stalk-like extension of lorica **15**

8. Stalk-like process formed from extension of lorica or pellicle (Fig.
   192A) **9**
   Body attached to substratum by stalk proper (Fig. 192B – Note
   stalk may be enclosed within lorica, Fig. 192C) **32**

9. Lorica difficult to distinguish, in form of a pellicle, without distinct
   apical aperture **10**
   Lorica distinct, cup-like, with definite open apical aperture **11**

10. Spherical highly contractile body. Capitate tentacles arise non-
    apically from single partly laterally displaced fascicle ....
    .... *Tokophryella* (p. 340)
    Oval non-contractile body. Capitate tentacles arise from all over
    apical region .... *Peridiscophrya* (p. 356)

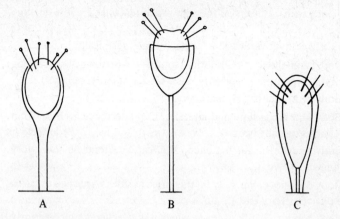

Fig. 192. Some methods of attachment in suctoria. A, An example of the extension of a lorica which in the key is not considered to be a proper stalk. B, Example of a stalk proper. C, Example of a stalk proper when enclosed by a lorica.

11. Lorica with vertical slits through which tentacles emerge laterally
.... *Metacineta* (p. 310)
Lorica without vertical slits, tentacles emerge apically **12**

12. Body contained within lorica so that only tentacles protrude
beyond aperture **13**
Body protrudes from lorica aperture **14**

13. Body and lorica flattened laterally. Capitate tentacles arise from
fascicles (usually two) .... *Periacineta* (p. 356)
Body and lorica not flattened. Non-capitate tentacles few, very
long and active .... *Urnula* (p. 316)

14. Body does not completely fill the lorica at sides. Attached to lorica
only at posterior end .... *Loricophrya* (p. 334)
Body completely fills lorica to which it is attached both posteriorly
and laterally .... *Caracatharina* (p. 344)

15. Body trumpet-shaped .... *Spelaeophrya* (p. 314)
Body not trumpet-shaped **16**

16. Body may be branched but always with arms or long processes
present (Fig. 193A–H) **17**
Body compact never branched, never with arms or long processes **24**

17. Macronucleus elongate and usually branched (Fig. 193A–D,F) .. **18**
Macronucleus rounded, compact never elongate never branched **20**

18. The arms or processes which project from the central body mass
do not branch (Fig. 193C,F) **19**
The arms or processes which project from the central body mass or
attachment site are distinctly branched (Fig. 193A,B,D) **22**

19. Arms long and finger-like with clubbed ends (Fig. 193C) ....
.... *Baikalophrya* (p. 324)
Arms with stout bases, narrowing towards ends, never clubbed
(Fig. 193F) .... *Lernaeophrya* (p. 334)

20. Arms branched, tentacles pointed (Fig. 193E,H) **21**
Arms not branched, club-like with capitate tentacles (Fig. 193G)
.... *Stylophrya* (p. 364)

21. Body low discoid shape, not mounted on column (Fig. 193H) .....
.... *Dendrocometes* (p. 348)
Body erect conical shape, mounted on column (Fig. 193E) ....
.... *Cometodendron* (p. 346)

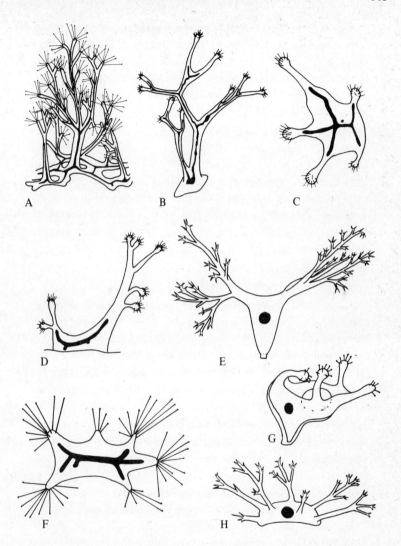

Fig. 193. Some suctoria with arm-like processes. A, *Dendrosoma*. B, *Gorgonosoma*. C, *Baikalophrya*. D, *Baikalodendron*. E, *Cometodendron*. F, *Lernaeophrya*. G, *Stylophrya*. H, *Dendrocometes*.

22. Attached part of body much branched, spreading over substratum, with branched erect arms leading from it. Tentacles long. (Fig. 193A).. *Dendrosoma* (p. 329)

    Attached to part of body compact not branched, not spreading but with branched erect arms leading from it. Tentacles short. (Fig. 193B,D) **23**

23. Macronucleus restricted to basin-shaped central body mass. Only arms held erect, much of body flattened on substratum. (Fig. 193D) .... *Baikalodendron* (p. 322)

    Macronucleus branches permeate throughout entire body and many of the branching arms. Much of body and arms held erect with only a small attachment area. (Fig. 193B) .... *Gorgonosoma* (p. 330)

24. Tentacles in fascicles **25**

    Tentacles not in fascicles **31**

25. With distinct lorica **26**

    Without lorica but pellicle may be present **27**

26. Lorica with six to eight vertical slits through which tentacles protrude (also has a stalk-like extension to lorica if viewed from side) .... *Metacineta* (p. 310)

    Lorica membranous, without vertical slits but with orifices, never with stalk-like extension to lorica .... *Solenophrya* (p. 336)

27. Tentacles pointed and regularly arranged in radial rows on apical surface... *Discosomatella* (p. 352)

    Tentacles capitate, in fascicles but not in rows **28**

28. Many tentacles in fascicles that are regularly arranged on body   **29**

    Few tentacles in fascicles that are irregularly arranged on body .... *Cyclophrya* (p. 346)

29. Body regularly circular in section. Never irregular nor with arch-like projections around body perimeter. With thick pellicle .... *Heliophrya* (p. 354)

    Body irregular in section, sometimes with arch-like projections around body perimeter. Pellicle thin **30**

30. Swarmer egg-shaped, produced by single or multiple endogenous budding. Often found on the peritrich *Epistylis* ..... *Trichophrya* (p. 342)

    Swarmer lentil-shaped, produced by simple endogenous budding. Fascicles on arch-shaped extensions of body. Never on *Epistylis* .... *Platophrya* (p. 352)

43. Conical body elongate and extended posteriorly to form a stalk-like projection terminating in short stalk. Lorica always close fitting and is completely filled with cytoplasm .... *Acinetides* (p. 320)

Conical body not elongate and extended posteriorly. Lorica may or may not be closely fitting and body is not always attached to lorica posteriorly .... *Acineta* (p. 318)

44. Few long, highly retractile tentacles **45**

Tentacles not retractile **49**

45. Single apical retractile tentacle **46**

More than one retractile tentacle **47**

46. One type of tentacle (highly mobile), without lorica   *Rhyncheta* (p. 336)

Usually single (one to six) retractile tentacle but also several smaller non-retractile capitate tentacles present, with lorica ....
.... *Acinetopsis* (p. 320)

47. Without lorica but with pellicle, single type of tentacle present ... **48**

With lorica, one to six retractile tentacles present plus several smaller capitate non-contractile ones .... *Acinetopsis* (p. 320)

48. Ellipsoid body, curved along longitudinal axis. Tentacles retractile but never expandable at tips. Macronucleus elongate ....
.... *Rhynchophria* (p. 362)

Spherical body, not curved. Tentacles both retractile and expandable at tips. Macronucleus rounded .... *Choanophrya* (p. 326)

49. Definite lorica with distinct apical aperture **50**

No lorica, two genera have pellicle, no apical aperture **51**

50. Much of body protrudes from lorica, aperture without collar ......
.... *Paracineta* (p. 312)

Body only just protrudes from lorica, aperture with collar ....
.... *Lecanodiscus* (p. 332)

51. Body rounded **52**

Body elongate .... *Discophrya* (p. 352)

52. Few tentacles which are thickened at tips .... *Suctorella* (p. 338)

Many tentacles which may be capitate or pointed but never thickened at tips **53**

Order SUCTORIDA

With the characteristics of the above subclass.

Suborder EXOGENINA

All members of this suborder produce exogenous buds, which usually are produced singly but multiple budding is known in certain species. The migratory larval forms are typically either large with complicated ventral ciliature or elongate where in some they may be practically devoid of cilia and incapable of swimming. These elongate larvae move in a worm-like manner.

Canellana Jankowski, 1967
(Fig. 194)

Description. Body an inverted conical shape lying within a distinct and easily distinguishable cup-like lorica. The latter is flattened, borne upon a stalk and there is an apical slit-like aperture although neither body nor tentacles protrude beyond the aperture. The short capitate tentacles are arranged in two fascicles. Macronucleus ovoid.

This single species genus is found in Lake Baikal growing upon the gammarid crustacean *Axelboekia carpenteri*. It was originally described by Swarczewsky (1928d) as *Thecacineta baikalica* and erected as a single species genus by Jankowski (1967a). *Canellana* is most easily confused with *Acineta* (p. 318).

Description of species. Swarczewsky (1928d).

Dendrosomides Collin, 1906
(Fig. 195)

Description. Erect branched body mounted on a short stalk. Short tentacles arise from distinct fascicles either on the ends of arms or at intervals along them. Macronucleus branched and reproduction is by exogenous budding to produce a worm-like swarmer. Freshwater species has several contractile vacuoles.

Single freshwater and single marine species described to date. Most easily confused with *Gorgonosoma* (p. 330) which does not have a stalk and also with *Dendrosoma* (p. 329) which has neither stalk nor erect habit.

Description of species. Gajewskaja (1933).

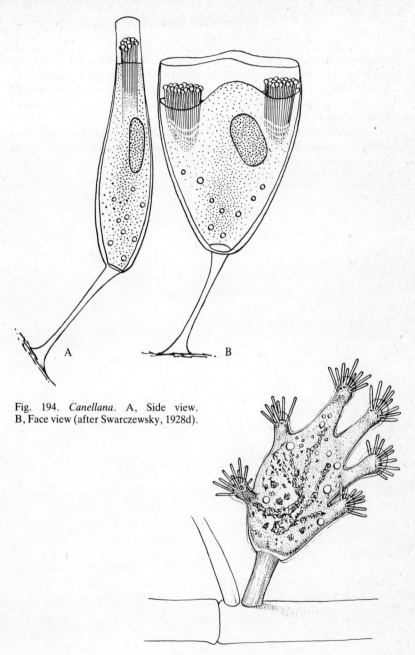

Fig. 194. *Canellana*. A, Side view.
B, Face view (after Swarczewsky, 1928d).

Fig. 195. *Dendrosomides* (after Gajewskaja, 1933).

Metacineta Bütschli, 1889
(Fig. 196)

Description. Body spherical to pyriform lying centrally within a distinct lorica which extends posteriorly to form a stalk-like region of variable length. The lorica is characteristically pierced by five, six or eight vertical slits through which the capitate tentacles protrude. The macronucleus is rounded and there are one or two contractile vacuoles.

One species *Metacineta pentagonalis* Nozawa, 1939, which has a very short stalk-like extension to the lorica could be easily confused with the genus *Solenophrya* (p. 336) if viewed from above.

Key to species. Nozawa (1939).

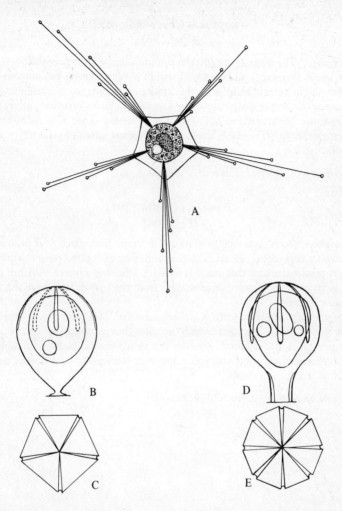

Fig. 196. *Metacineta*. A, Apical view. B,C, Lateral and apical views of species with five slits in lorica. D,E, Lateral and apical views of species with eight slits in lorica (after Nozawa, 1939).

Mucophrya Gajewskaja, 1928
(Fig. 197)

Description. The body is irregular in outline, lumpy, of a grey-yellow colour (100 μm in diameter) and enclosed within a very transparent mucous coat (130μm in diameter). There is neither stalk nor other point of attachment and the animal is completely pelagic. The capitate tentacles arise irregularly from all over the body surface. Macronucleus rounded. Cyst is a characteristic parachute shape. This is a single species genus originally found in Lake Baikal.

Description of species. Gajewskaja (1928).

Paracineta Collin, 1911
(Fig. 198)

Description. Body spherical to ellipsoid in shape with much of it protruding out from a cup-shaped lorica. The lorica may be elongated posteriorly but it always terminates in a true stalk. Tentacles arise and radiate out from all of that portion of the body that protrudes from the lorica. Reproduction is by exogenous buds.

Most easily confused with *Caracatharina* (p. 344) which does not protrude from the lorica and which by invaginative budding, produces a swarmer which has ciliation covering the entire ventral surface. In addition *Caracatharina*'s conjugation type represents anisogamy and partial conjugation of adult individuals.

Description of species. Collin (1912).

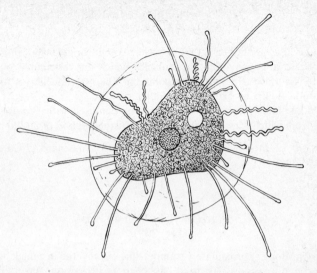

Fig. 197. *Mucophrya* (after Gajewskaja, 1928).

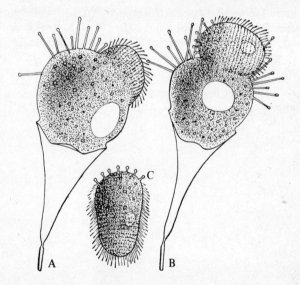

Fig. 198. *Paracineta*. A,B, Adults with developing exogenous buds. C, Embryo with developing tentacles (after Collin, 1912).

Podophrya Ehrenberg, 1838
(Fig. 199)

Description. Body spherical to ovoid borne on hollow rigid stalk. There is no lorica. There are many capitate tentacles which are distributed over the entire body surface. Encystment is common. The macronucleus is rounded and reproduction is by exogenous budding.

This a common genus that may be easily confused with *Prodiscophrya* (p. 360) from which it may be distinguished by the method of budding only.

Descriptions of species. Collin (1912), Kahl (1934), Penard (1920), Sand (1899–1901).

Spelaeophrya Stammer, 1935
(Fig. 200)

Description. Body approximately trumpet-shaped, covered in a thick pellicle. Posterior end of body narrowed, attached to freshwater invertebrates, particularly shrimps. Sometimes has a rudimentary lorica at its base which is an extension of the pellicle. Tentacles are restricted to the perimeter of the wide apical region. They may be capitate and of medium length (Stammer, 1935) or short and non-contractile (Nozawa, 1938). Macronucleus is band-like and reproduction is by exogenous budding producing vermiform larvae. Recent work by Matthes and Plachter (1978) has described the development of these vermiform unciliated larvae which actively move in a leech-like manner.

Descriptions of species. Matthes and Plachter (1978), Nozawa (1938), Stammer (1935).

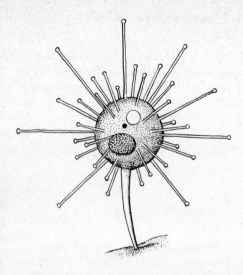

Fig. 199. *Podophrya* (after Bütschli,
              1887–89).

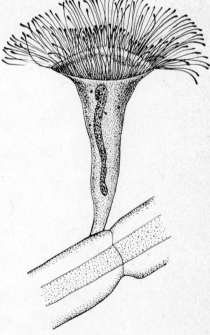

Fig. 200. *Spelaeophrya* (after Stammer,
                    1935).

Sphaerophrya Claparède and Lachmann, 1859
(Fig. 201)

Description. Spherical body without lorica and stalk, floating free on the water or attached by its tentacles to other organisms. Tentacles not in fascicles but distributed evenly over entire body surface. Macronucleus round and there is a single contractile vacuole. There are several species, some are free-living but some are parasites. Reproduction is by binary fission or by exogenous budding.

Descriptions of species. Grandori and Grandori (1934), Penard (1920).

Urnula Claparède and Lachmann, 1859
(Fig. 202)

Description. Body approximately oval situated within, but not filling, an irregularly cup-shaped lorica. The posterior region of the lorica is constricted into a stalk-like attachment point which is usually bent to one side. There is an apical rounded or triangular aperture through which one or two (but up to five) long active non-capitate tentacles protrude. The apical end of the lorica is narrower than the central region. The nucleus is oval and centrally situated. There are one or more contractile vacuoles. The organism is usually reported to be found growing on the stalks of *Epistylis plicatilis*.

General biology. Kormos (1958), Kormos and Kormos (1958b), Penard (1920).

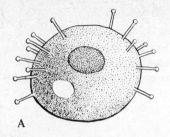

A

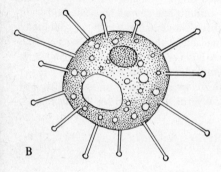

B

Fig. 201. *Sphaerophrya*. A, After Claparède and Lachmann (1859). B, After Penard (1920).

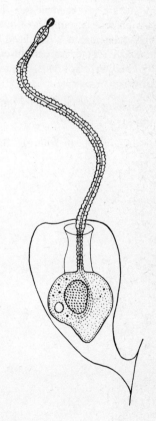

Fig. 202. *Urnula* (after K. Kormos, 1958).

Suborder ENDOGENINA

All members of this suborder produce one or more endogenous buds completely internally inside a brood pouch. The larvae gain cilia and become motile inside the brood pouch before emergence through a birth pore. The larval forms are typically small with encircling band or bands of cilia.

Acineta Ehrenberg, 1833
(Fig. 203)

Description. Body inverted conical shape within a cup-like lorica which in some species may be close fitting and difficult to observe. The body may or may not completely fill the lorica which is flattened and borne upon a stalk. The lorica opens apically usually as a dumb-bell shaped slit. The body and tentacles protrude through this aperture. The tentacles which are commonly capitate are arranged in two (but may be in three) distinct fascicles. The macronucleus is rounded to ovoid. A larva which has a ciliated band or is completely ciliated is produced by simple endogenous budding.

Acineta is most easily confused with *Acinetides* (p. 320), *Canellana* (p. 308) and *Tokophrya* (p. 340). There are several species and the genus may be freshwater or marine.

Descriptions of species. Collin (1912), Kahl (1934), Penard (1920), Swarczewsky (1928a).

Budding and metamorphosis. Bardele (1970).

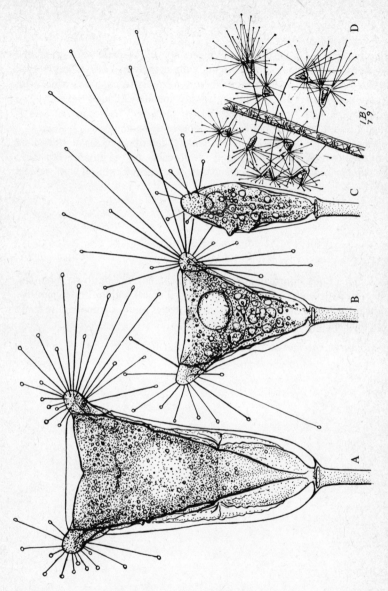

Fig. 203. *Acineta*. A,B, Face views. C, Side view. D, Many suctoria growing on an algal filament (all drawn from life).

Acinetides Swarczewsky, 1928
(Fig. 204)

Description. Body is an inverted elongate conical shape which is extended posteriorly like a stalk but is filled with cytoplasm. This latter region terminates in a short stalk proper. There is a closely fitting lorica which with the exception of the two tentacle-regions is completely filled by the body. The tentacles are short and arranged in two fascicles. The macronucleus is elongate and a larva is produced by endogenous budding.

This is a single species genus originally described in Lake Baikal by Swarczewsky (1928d), who found it growing on gammarid crustacea (*Poeckilogammarus pictus*). *Acinetides* is most easily confused with *Acineta* (p. 318).

Acinetopsis Robin, 1879
(Fig. 205)

Description. Body approximately conical shape having one to six large and highly retractile tentacles situated in a central apical region. Lorica cup-like and distinct, borne upon a stalk. Major tentacles often surrounded by many very much smaller tentacles.

Description of species. Swarczewsky (1928d).

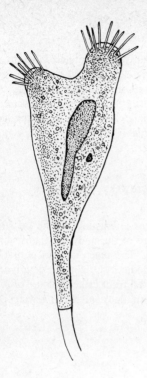

Fig. 204. *Acinetides* (after Swarczewsky, 1928d).

Fig. 205. *Acinetopsis* with one tentacle (after Swarczewsky, 1928d).

Astrophrya Awerinzew, 1904
(Fig. 206)

Description. Stellate-shaped body with central mass pulled out into eight elongate projections which terminate in a fascicle of capitate tentacles. The body is covered in sand grains and other objects except for the areas immediately adjacent to the tentacles. There is no stalk nor any other point of attachment and the animal is planktonic. A distinctive single species genus. Sometimes wrongly spelt *Asterophrya*.

Baikalodendron Swarczewsky, 1928
(Fig. 207)

Description. Irregularly shaped body forms encrusting growth on surface of gammarid crustaceans. Around the edge of the body there arise finger-like projections which bear clubbed ends. The macronucleus is branched and reproduction is by endogenous budding. Originally described by Swarczewsky (1928a) from material collected from Lake Baikal.

Most easily confused with *Baikalophrya* (p. 324) and *Stylophrya* (p. 364).

Description of species. Swarczewsky (1928d).

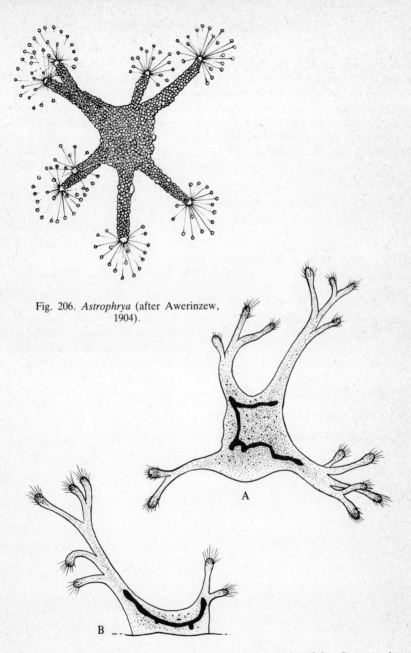

Fig. 206. *Astrophrya* (after Awerinzew, 1904).

Fig. 207. *Baikalodendron*. A, Apical view. B, Lateral view (after Swarczewsky, 1928d).

Baikalophrya Swarczewsky, 1928
(Fig. 208)

Description. Body is an irregular shape forming an encrusting growth on gammarid crustacea such as *Acanthogammarus*. Unbranched finger-like outgrowths project from around the edge of the body, their ends are clubbed and terminate in tentacles. The macronucleus is band-like and branched. Reproduction by endogenous budding. Originally described from Lake Baikal, not yet reported elsewhere.

Most easily confused with *Stylophrya* (p. 364) which is mounted upon a column-like support and with *Baikalodendron* (p. 322) whose arms are branched.

Description of species. Swarczewsky (1928a).

Brachyosoma Batisse, 1975
(Fig. 209)

Hallezia Sand, 1895

Description. Body spherical with a small adhesive protoplasmic stalk-like projection by which the animal attaches itself to the substratum. Without a lorica. Capitate tentacles in two, three or four fascicles. Two contractile vacuoles present. Macronucleus ovoid and approximately centrally placed.

The genus is most easily confused with *Tokophrya* (p. 340) which has a proper stalk not containing protoplasm.

Description of species. Stokes (1885b).

Synonymy and species list. Collin (1912), Batisse (1975b).

Fig. 208. *Baikalophrya*
(after Swarczewsky, 1928a).

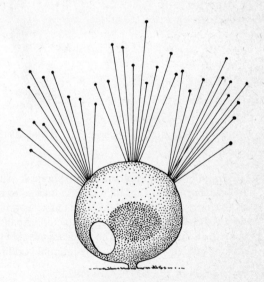

Fig. 209. *Brachyosoma* (after Stokes,
1885a).

Choanophrya Hartog, 1902

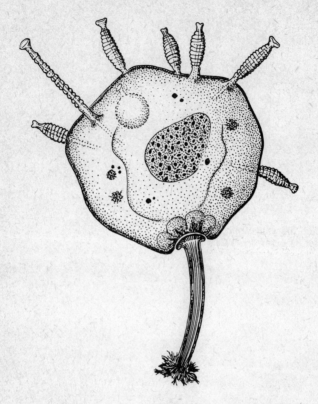

Fig. 210 *Choanophrya* (after Hartog, 1902 and Farkas, 1924).

Description. Body approximately spherical, without lorica but with thick pellicle, borne upon a stalk which is attached to freshwater crustaceans (such as *Cyclops*). There are 12–15 large tentacles which emerge from all over the body surface; these are wide, contractile, expansible at their distal ends and contain an internal canal. The tentacles are able to engulf large particles by a sucking action. Reproduction by internal budding, macronucleus oval.

Description of species. Penard (1920).

Tentacles and feeding. Farkas (1924).

Cryptophrya Jankowski, 1973

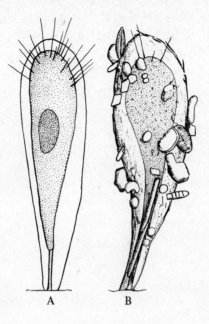

A   B

Fig. 211. *Cryptophrya.* A, Longitudinal section. B, External appearance (after Swarczewsky, 1928c).

Description. Body elongated conical shape mounted on a short, narrow but distinct stalk. The entire body and stalk surrounded by a distinct balloon-like lorica which tends to become covered in sand grains and other debris. Pointed tentacles emerge from all over the anterior end of the body and protrude through the lorica. The macronucleus is ovoid and reproduction is by simple endogenous budding.

The genus was originally described by Swarczewsky (1928c) as *Discophrya obtecta* found growing on freshwater gammarid crustaceans in Lake Baikal. More recently Jankowski (1973a) thought the animal sufficiently different to erect it as a new single species genus.

Dactylostoma Jankowski, 1967

Fig. 212. *Dactylosoma* (after Gajewskaja, 1929).

Description. Approximately ellipsoid shaped body mounted on a short, stout longitudinally striated stalk. The body lies within a tightly fitting pellicle-like lorica. Twelve to fifteen tentacles emerge apically, they are short, stout non-capitate, non-contractile and contain an internal tube.

The genus was first described as *Dactylophrya collini* Gajewskaja, 1929 and was recently (Jankowski, 1967a) transferred into a new single species genus *Dactylostoma*. Found growing on antennae of crustaceans.

Dendrosoma Ehrenberg, 1838

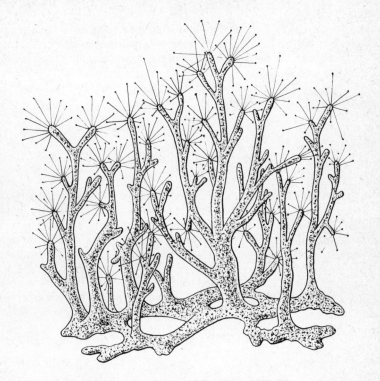

Fig. 213. *Dendrosoma* (after Kent, 1880–82).

Description. Large (1–2.5 mm high) irregularly spreading branched dendritic body with clusters of capitate tentacles located at ends of branches. Neither stalk nor lorica present although it has a tendency to collect a covering of adherent sand grains and plant detritus. A large proportion of the body is attached directly to submerged vegetation and/or invertebrate animals over which it spreads and branches giving rise to erect branched arms at intervals. There are several contractile vacuoles. The macronucleus is band-like and branched with numerous micronuclei. Reproduction is by endogenous budding and the exogenous buds often illustrated were shown by Hickson and Wadsworth (1909) to be those of the epizooic suctorian *Urnula* (p. 316).

The genus is most easily confused with *Dendrosomides* (p. 308) which is erect with a stalk and with *Gorgonosoma* (p. 330) which is erect but without a stalk.

Descriptions of species. Gönnert (1935), Hickson and Wadsworth (1909).

Taxonomy. Gönnert (1935).

Erastophrya Fauré-Fremiet, 1944
(Fig. 214)

Description. Body pyriform shape with posterior end drawn out into a pair of claspers which are used to grip the base of a peritrichous ciliate. Capitate tentacles present, distributed over most of the body. Macronucleus spherical or elongate with reproduction by simple endogenous budding.

A single species genus which is ectocommensal on the peritrich *Apiosoma piscicola* (Fauré-Fremiet, 1944a).

Gorgonosoma Swarczewsky, 1928
(Fig. 215)

Description. Large irregularly branched dendritic body with clusters of tentacles located at the clubbed ends of branches. Neither stalk nor lorica present. The majority of the body is erect although its non-branching base spreads to a limited extent. The macronucleus is band-like and branched with several micronuclei. Reproduction is by endogenous budding. Found in Lake Baikal growing on gammarid crustaceans.

The genus is most easily confused with *Dendrosoma* (p. 329) which has a branched spreading habit and with *Dendrosomides* (p. 308) which is erect but has a stalk.

Description of species. Swarczewsky (1928a).

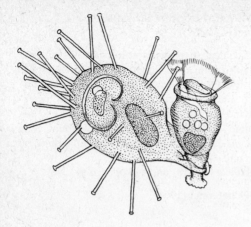

Fig. 214. *Erastophrya* growing on the peritrich *Apiosoma* (after Fauré-Fremiet, 1944).

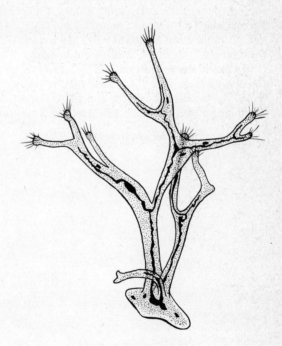

Fig. 215. *Gorgonosoma* (after Swarczewsky, 1928a).

Hypophrya López-Ochoterena, 1964
(Fig. 216)

Description. Body an inverted conical shape without a lorica but within a thick resistant pellicle. Zooid borne upon a long longitudinally striated stalk. Capitate tentacles arranged in five fascicles, four situated apically and one encircling the stalk base. Two anterior contractile vacuoles. The macronucleus is ovoid or cylindrical and rounded embryos are produced endogenously. Single species genus found in Mexico. The organism was found as an epibiont of the peritrich *Epistylis plicatilis* which itself was epibiontic on shells of the molluscs *Limnaea* and *Physa*.

Most easily confused with *Acineta* (p. 318) and *Tokophrya* (p. 340).

Description of species. López-Ochoterena (1964).

Lecanodiscus Jankowski, 1973
(Fig. 217)

Description. Body ovoid to elongated cone shape, within a lorica which has a circular apical opening. The lorica forms a distinct collar around the aperture through which the body just protrudes and tentacles completely project. The lorica is borne upon a striated stalk. Macronucleus is ovoid to elongate and reproduction is by simple endogenous budding.

This genus was recently created by Jankowski (1973a) for the two species *Discophrya longa* and *Thecacineta robusta* originally described from Lake Baikal by Swarczewsky (1928c and 1928d respectively). Both species were found growing on the crustacean *Axelboekia carpenteri*.

Descriptions of species. Swarczewsky (1928c,d).

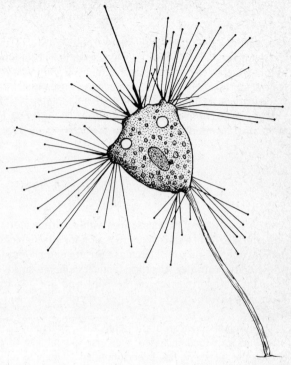

Fig. 216. *Hypophrya* (after López-
Ochoterena, 1964).

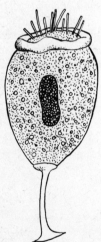

Fig. 217. *Lecanodiscus*
(after Swarczewsky, 1928c).

Lernaeophrya Pérez, 1903
(Fig. 218)

Description. Large irregularly shaped body covered with numerous, usually short, finger-like prolongations which terminally bear long capitate tentacles. Found attached to both animals and plants in various ways. There are several contractile vacuoles. The macronucleus is usually band-like and branched. Reproduction by multiple endogenous budding.

Most easily confused with *Trichophrya* (p. 342) and *Platophrya* (p. 358). Sometimes covered with the suctorian ciliate *Urnula* (K. Kormos, 1958).

Loricophrya Matthes, 1956
(Fig. 219)

Description. Body of various shapes within a distinct lorica which is extended posteriorly to form a stalk-like projection for attachment purposes. There is a definite free margin to the circular aperture of the lorica through which some of the body protrudes. Capitate tentacles emerge mostly but not exclusively from the anterior surface of the body and are not arranged in fascicles. Macronucleus rounded, mode of reproduction and swarmer not yet described.

The original diagnosis of the genus *Loricophrya* by Matthes (1950) and list of species provided included a diverse assemblage of organisms. For this reason the diagnosis has been amended to include only those which have generic characters closely similar to those of the genotype *Loricophrya parva* (Shulz, 1931) Matthes, 1956. The three freshwater species described by Maskell (1886, 1887) are not considered by the present author to belong to the genus *Loricophrya* as suggested by Matthes (1956) since they possess true stalks which are not just extensions of the lorica and further possess tentacles arranged in definite fascicles. The genus is most easily confused with *Metacineta* (p. 310) and *Urnula* (p. 316).

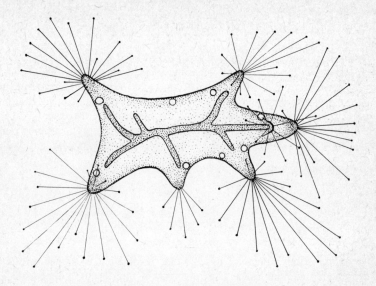

Fig. 218. *Lernaeophrya* (after K. Kormos, 1958).

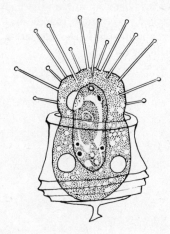

Fig. 219. *Loricophrya* (after Matthes, 1956).

Rhyncheta Zenka, 1866
(Fig. 220)

Description. Elongate or irregularly spherical body without lorica, rounded apically but flattened terminally near the stalk. A single long (500 μm when extended) high retractile (50 μm when contracted) tentacle emerges from the anterior end although other tentacles under development or being resorbed may also be present. Body attached to the freshwater crustacean (*Cyclops*) by a short (4 μm long) stalk. Macronucleus rounded, reproduction by rapid or multiple endogenous budding to produce barrel-shaped swarmers.

Description of species. Hitchen and Butler (1972).

Solenophrya Claparède and Lachmann, 1859
(Fig. 221)

Description. Body spherical to ellipsoid, completely contained within a membranous globular lorica. The lorica may be translucent or opaque due to the presence of foreign particles, it may be closed, open apically or pierced by small orifices near to the apex. The animal is attached by its lorica direct to the substratum without an intervening stalk. There are one to six fascicles of capitate tentacles which radiate out around the body. There are one to several contractile vacuoles, a centrally placed oval macronucleus and reproduction is by endogenous budding. The oval larvae produced have cilia in oblique rows around the body.

Descriptions of species. Hull (1954), Penard (1920).

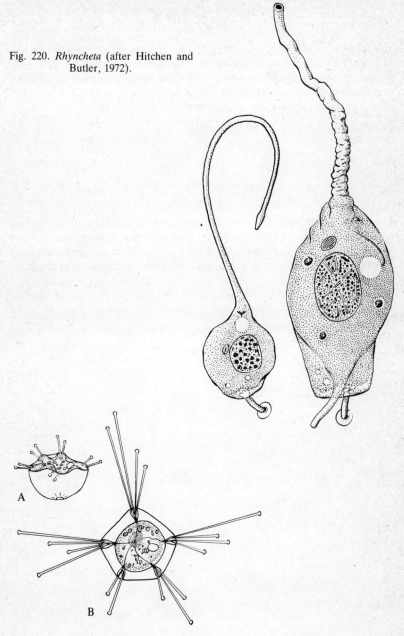

Fig. 220. *Rhyncheta* (after Hitchen and Butler, 1972).

Fig. 221. *Solenophrya.* A, Lateral view. B, Apical view (after Hull, 1954).

Staurophrya Zacharias, 1893
(Fig. 222)

Description. Rounded body bearing six regularly placed short arm-like extensions each of which bear tentacles. Macronucleus is rounded, there may be one or two contractile vacuoles. Reproduction is by simple endogenous bud formation. Without stalk, without lorica.

Staurophrya is most easily confused with *Astrophrya* (p. 322).

Description of species. Zacharias (1893).

Suctorella Frenzel, 1891
(Fig. 223)

Description. Body spherical extended posteriorly as a small cone from which a fine stalk arises. There is no lorica. Tentacles few, arising from all over body and they have distinctly thickened ends. There are two contractile vacuoles, the macronucleus is spherical and reproduction is by endogenous budding.

Originally described by Frenzel (1891) who found it in a muddy pool in Argentina. Sand (1899–1901) and Collin (1912) both consider it to be synonymous with *Tokophrya* (p. 340).

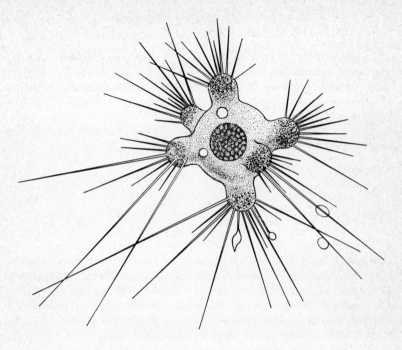

Fig. 222. *Staurophrya* (after Zacharias, 1893).

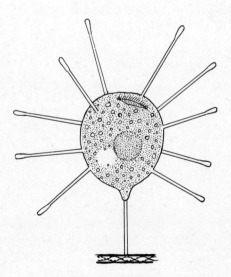

Fig. 223. *Suctorella* (after Sand, 1899–1901).

Tokophrya Bütschli, 1889
(Fig. 224)

Description. Body inverted pyriform or conical in shape entirely without a lorica and rounded not flattened in cross-section. Borne upon a non-rigid stalk. Capitate tentacles are arranged in two to four fascicles, all on the anterior surface. Macronucleus rounded. Ciliated larva produced by simple endogenous budding.

Most easily confused with *Acineta* (p. 318) and *Acinetides* (p. 320), the former being flattened and both being loricate.

Description of species. Collin (1912).

Tokophryella Jankowski, 1973
(Fig. 225)

Description. Rounded, very contractile body borne upon a short peduncle that appears to be an extension of the body tegument. The capitate tentacles arise from a single anterior fascicle that is positioned to one side of the body. The macronucleus is spherical and reproduction is by simple endogenous budding. This is a single species genus described by Claparède and Lachmann (1961) as *Podophrya carchesii* who found it growing on the contractile stalk of the peritrich *Carchesium polypinum*, recently Jankowski (1973a) thought it sufficiently different to warrant erecting it as a new genus.

Description of species. Claparède and Lachmann (1861).

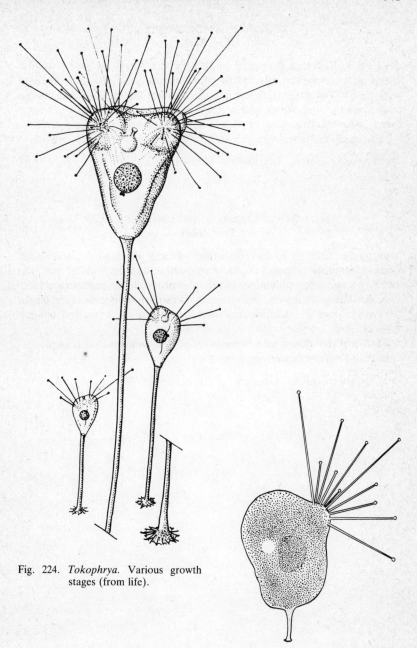

Fig. 224. *Tokophrya*. Various growth
stages (from life).

Fig. 225. *Tokophryella* (after Claparède
and Lachmann, 1861).

Tokophryopsis Swarczewsky, 1928
(Fig. 226)

Description. Inverted pyramidal body without a lorica and not flattened. Borne upon a long stalk attached to gammarid crustacea (*Poeckilogammarus sukaczewi*). Tentacles arranged in three distinct fascicles on the anterior surface of the body from within three crown-like protoplasmic protuberances. Macronucleus elongate. Single species genus found in Lake Baikal.

Most easily confused with *Acineta* (p. 318) and *Tokophrya* (p. 340).

Description of species. Swarczewsky (1928d).

Trichophrya Claparède and Lachmann, 1859
(Fig. 227)

Description. Body irregularly spherical, directly attached to substratum. Capitate tentacles long and regularly arranged in fascicles. Neither stalk nor lorica present. Reproduction by simple or multiple endogenous budding producing an egg-shaped swarmer. Often reported growing upon the peritrich *Epistylis*. Macronucleus ovoid, elongate or branched but not coiled.

Most easily confused with *Platophrya* (p. 358) with which it is sometimes considered to be synonymous.

Descriptions of species. Gönnert (1935), Penard (1920).

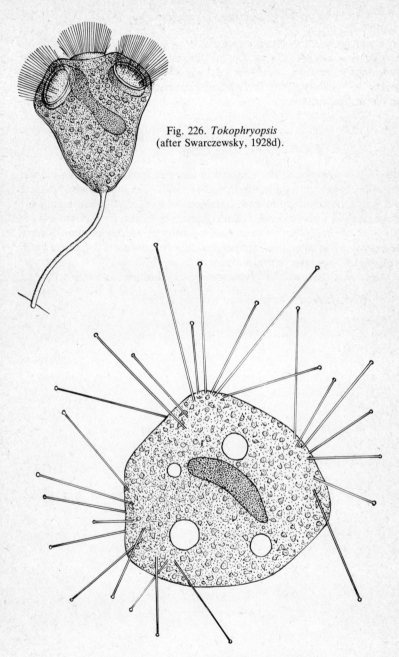

Fig. 226. *Tokophryopsis* (after Swarczewsky, 1928d).

Fig. 227. *Trichophrya* (after Gönnert, 1935).

Suborder EVAGINOGENINA

All members of this suborder reproduce by evaginative budding with the development of a single larva. The bud begins in a typical endogenous brood pouch but is everted and only then does it develop cilia and become motile. Larvae are typically ellipsoidal, flattened and bear distinctive ciliary patterns on the ventral surface.

Caracatharina Kormos, 1960
(Fig. 228)

Description. Body elongate, ellipsoid in shape of which only the apical part protrudes from out of the distinct cup-like lorica. The lorica is also elongate and has a posterior prolonged extension to form a stalk-like region for attachment via a broad circular plate. Capitate tentacles emerge from all over that part of the body that protrudes beyond the lorical aperture. Reproduction is by invaginative internal budding to produce a swarmer which is ciliated all over the ventral surface.

Most easily confused with *Paracineta* (p. 312) which has a true stalk.

Description of species. Kormos and Kormos (1958a).

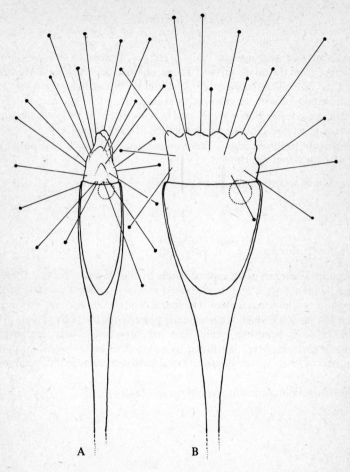

Fig. 228. *Caracatharina*. A, Side view. B, Face view (after Kormos and Kormos, 1958a).

Cometodendron Swarczewsky, 1928
(Fig. 229)

Description. Elongate upright body bearing a number of branched arms which arise from the upper surface. There is no stalk and no lorica. The body is attached directly to freshwater gammarids by a foot-like extension. The arms are equipped with short retractile pointed tentacles. The macronucleus is spherical and reproduction is reported to be by simple endogenous budding. Originally described growing on gammarids in Lake Baikal.

Most easily confused with *Dendrocometes* (p. 348) which has a rounded body without a foot attachment.

Description of species. Swarczewsky (1928b).

Cyclophrya Gönnert, 1935
(Fig. 230)

Description. Flattened body approximately circular in cross-section. There is neither lorica nor stalk and the flattened animal is attached directly to the substratum by a large basal plate. The sparse capitate tentacles are situated in three to six fascicles which are irregularly placed over the body surface. The macronucleus is band-like, coiled but not branched. There are several contractile vacuoles. Reproduction is by endogenous bud formation with the production of an elongate cylindrical larval form with many longitudinal rows of cilia.

Most easily confused with *Heliophrya* (p. 354).

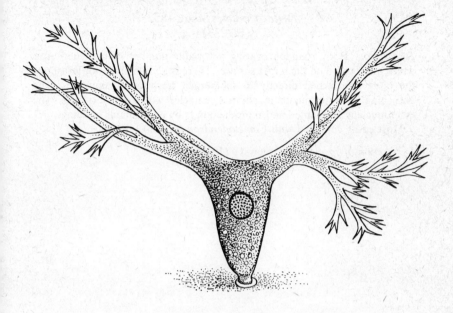

Fig. 229. *Cometodendron* (after Swarczewsky, 1928b).

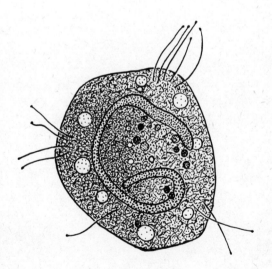

Fig. 230. *Cyclophrya* (after Gönnert, 1935).

Dendrocometes Stein, 1852
(Fig. 231)

Description. Body rounded bearing a variable number of branched arms which are located on the upper surface. There is neither stalk nor lorica and the body is attached directly to submerged freshwater gammarids. The branching arms terminate in pointed tentacles which are retractile. The macronucleus is rounded and reproduction is by endogenous budding.

Most easily confused with *Cometodendron* (p. 346).

Description of species. Swarczewsky (1928b).

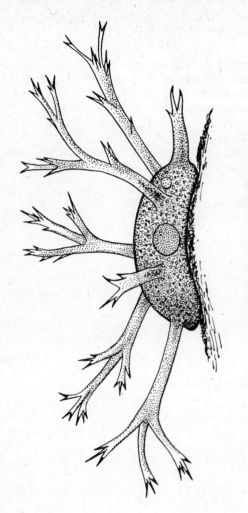

Fig. 231. *Dendrocometes* (after Swarczewsky, 1928b).

Dendrocometides Swarzewsky, 1928
(Fig. 232)

Description. Body hemispherical in shape with flattened side attached directly to substratum. There is neither lorica nor stalk present. The tentacles are slender and pointed and arise from the dorsal surface. Some tentacles are branched, others are simple. The macronucleus is rounded and the animal reproduces by endogenous budding. Found growing on freshwater gammarids in Lake Baikal.

Most easily confused with *Stylocometes* (p. 364) which has unbranched tentacles and *Discosomatella* (p. 352) which has unbranched tentacles arranged in fascicles.

Description of species. Swarczewsky (1928b).

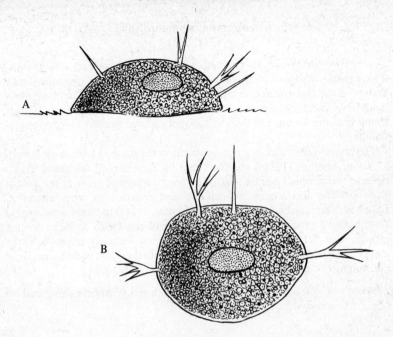

Fig. 232. *Dendrocometides*. A, Lateral view. B, Apical view (after Swarczewsky, 1928b).

Discophrya Lachmann, 1859
(Fig. 233)

Description. Elongate body covered in a pellicle, mounted on a short stalk. The capitate tentacles may be either evenly distributed over the apical surface of the body or in several fascicles all over the body. There are many contractile vacuoles each with canals leading to the body surface. Usually found growing on aquatic insects. Ciliated embryo formed by endogenous budding.

The above is the older description strictly in the sense of Lachmann (1859). However Matthes (1954a) considers that the members of the genus have a definite lorica which can be absent in young forms and grow as the animal matures. The lorica may be cup-like and rounded in cross-section or flattened. The attachment is also said (Matthes, 1954a) to be variable ranging from being a posterior stalk-like extension of the lorica to being a stalk proper. The flattened embryo which is ciliated ventrally is produced by invaginative budding. The latter definition would therefore include species of the genera *Periacineta* (p. 356) and *Peridiscophrya* (p. 356).

Descriptions of species. Matthes (1953, 1954a,b,e,d,f), Matthes and Stiebler (1970), Swarczewsky (1928c).

Discosomatella Corliss, 1960
(Fig. 234)

Discosoma Swarczewsky, 1928

Description. Discoid shaped body whose base attaches direct to the substratum without stalk. No lorica. There are eight to ten rows of short pointed tentacles regularly arranged radially around the apical surface of the animals. There are four to six tentacles in each row. The centrally-placed macronucleus is spherical. Reproduction by simple endogenous budding.

Most easily confused with *Dendrocometides* (p. 350) which has branched pointed tentacles that are not in fascicles.

Description of species. Swarczewsky (1928b).

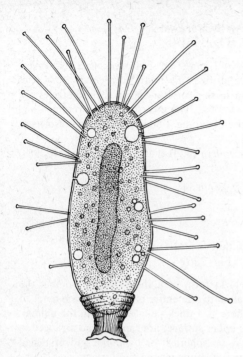

Fig. 233. *Discophrya* (after Collin, 1912).

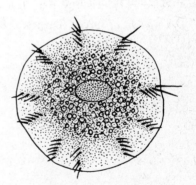

Fig. 234. *Discosomatella*
(after Swarczewsky, 1928b).

Echinophrya Swarczewsky, 1928
(Fig. 235)

Description. Body spherical to ovoid, borne on short stout stalk. No lorica. The tentacles may emerge from all over the body surface not in fascicles although they are mainly found in the anterior region. The tentacles are distinctive being pointed and spine-like. The macronucleus is rounded and reproduction is by multiple internal budding. Found growing on crustaceans in Lake Baikal.

Most easily confused with *Podophrya* (p. 314) and *Prodiscophrya* (p. 360) which have capitate tentacles.

Description of species. Swarczewsky (1928c).

Heliophrya Saedeleer and Tellier, 1930
(Fig. 236)

Description. Body approximately in form of a short cylinder so that the animal is attached to the substratum by the entire basal area which tends to spread out. Characteristically there is a thick pellicle covering the animal. The tentacles are located on the upper surface, are capitate and arranged in fascicles. The macronucleus may be spherical or band-like and branched according to species. Reproduction by endogenous bud formation.

Most easily confused with *Cyclophrya* (p. 346) which has few tentacles arranged in regular fascicles.

Descriptions of species. Dragesco. Blanc-Brude and Skreb (1955), Lanners (1973), Matthes (1954e).

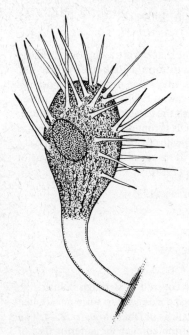

Fig. 235. *Echinophrya*
(after Swarczewsky, 1928c).

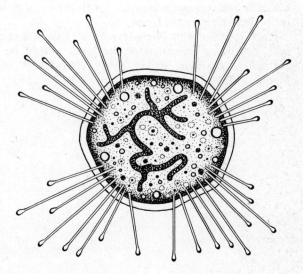

Fig. 236. *Heliophrya* (after Dragesco,
Blanc-Brude and Skreb, 1955).

Periacineta Collin, 1909
(Fig. 237)

Description. Body usually inverted conical shape with a cup-like elongate lorica whose base is extended posteriorly to form a stalk-like region. The adult animal is flattened and the lorica may have transverse markings anteriorly denoting lines of growth (Nozawa, 1938). There is an apical slit-like aperture in the lorica through which the capitate tentacles protrude, usually from two fascicles. The macronucleus is rounded and reproduction is by endogenous buds. Most easily confused with *Acineta* (p. 318) which has a distinct stalk at the base of the lorica. It is regarded as a synonym of *Discophrya* (p. 352) according to Matthes (1954e).

Descriptions of species. Collin (1912), Nozawa (1933).

Peridiscophrya Nozawa, 1938
(Fig. 238)

Description. Body cylindrical and entirely covered by a tough pellicle-like lorica. Posteriorly the lorica is constricted and extended to form a stalk-like region although it never possesses a true stalk as in *Discophrya* (p. 352). The capitate tentacles are restricted to and distributed over the anterior face of the animal. Small sand grains and other foreign objects often adhere to the pellicle. The macronucleus is band-shaped and irregularly branched. There are up to four contractile vacuoles. Originally found growing on the shell of the mollusc *Viviparus* in Japan.

Easily confused with *Discophrya* (p. 352) which is attached by a definite stalk not by an extension of the lorica. Could be synonymous with *Discophrya* according to the newer diagnosis of the latter genus by Matthes (1954e).

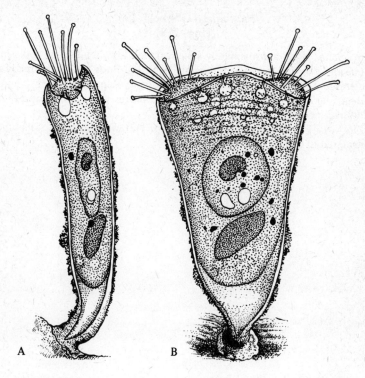

Fig. 237. *Periacineta*. A, Side view. B, Face view (after Collin, 1912).

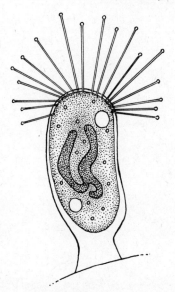

Fig. 238. *Peridiscophrya* (after Nozawa, 1938).

Platophrya Gönnert, 1935
(Fig. 239)

Description. Body approximately spherical with low arch-like protoplasmic extensions carrying fascicles of capitate tentacles. Body attached directly to substratum. Neither stalk nor lorica present. Reproduction by simple endogenous budding producing lentil-shaped swarmer. Macronucleus oval to band-like, not branched, not coiled.

Most easily confused with *Trichophrya* (p. 342) with which it is sometimes considered to be synonymous.

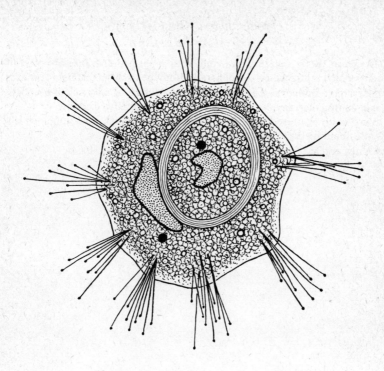

Fig. 239. *Platophrya* (after Gönnert, 1935).

Prodiscophrya Kormos, 1935
(Fig. 240)

Description. Body spherical to ovoid, with slight protuberance at posterior end to which is attached a slender stalk. Body bears 30–60 capitate tentacles which radiate from over the entire surface. No lorica. Macronucleus rounded, single contractile vacuole. Reproduction by invaginative budding.

Very easily confused with *Podophrya* (p. 314) which is morphologically identical and is distinguished by its exogenous budding.

Genus erected for the species *Podophrya collini* Root, 1914.

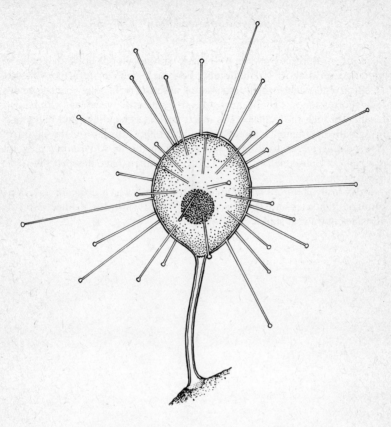

Fig. 240. *Prodiscophrya* (after Kormos, 1935).

Rhynchophria Collin, 1909
(Fig. 241)

Description. Body curved ellipsoid shape without a lorica but with a pellicle. Mounted on a short striated stalk. Few large mobile retractile tentacles at apical end which usually consist of a single main long tentacle and a few shorter ones. There are several contractile vacuoles with canals leading through the pellicle. The macronucleus is band-like and budding is invaginative. Originally found by Collin (1909) growing on the elytra of *Hydrophilus piceus* who spelt the generic name as *Rhynchophria* in his description. Subsequently Collin (1912) and others have misspelt the name as *Rhynchophrya*.

Most easily confused with *Discophrya* (p. 352) which does not have long retractile tentacles and *Peridiscophrya* (p. 356) which has neither retractile tentacles nor a real stalk.

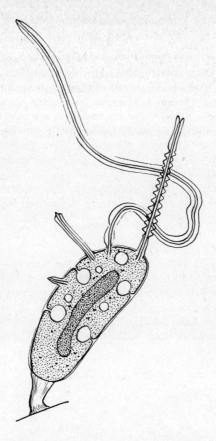

Fig. 241. *Rhynchophria* (after Collin, 1912).

Stylocometes Stein, 1867
(Fig. 242)

Description. Body rounded without lorica and without stalk, attached directly to the gills of submerged aquatic invertebrates such as *Asellus* and *Aphrydium*. There are about 20 unbranched non-retractile tentacles which arise from the dorsal surface. Macronucleus is elongate with two to three micronuclei.

Most easily confused with *Dendrocometides* (p. 350) which has branching tentacles.

Description of species. Skreb-Guilcher (1955).

Stylophrya Swarczewsky, 1928
(Fig. 243)

Description. Body basin or cup-shape with the posterior drawn out into a supporting column which is directly attached to gammarid crustaceans. Unbranched club-like arms project from around the perimeter of the body and these bear short terminal capitate tentacles. The macronucleus is rounded. Found in Lake Baikal.

Most easily confused with *Baikalodendron* (p. 322) and *Baikalophrya* (p. 324).

Description of species. Swarczewsky (1928a).

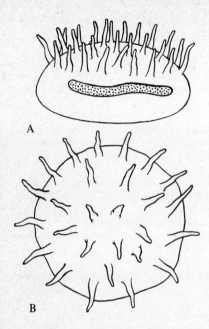

Fig. 242. *Stylocometes*. A, Lateral view.
B, Apical view (after Stein, 1867).

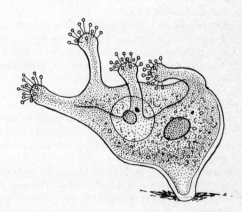

Fig. 243. *Stylophrya* (after Swarczewsky, 1928a).

Glossary of Special Terms

aboral At the end of the body opposite to the oral region.

adoral zone of membranelles A serial arrangement of three or more membranelles along the left side of the oral area. Commonly abbreviated to AZM.

apical At the extreme anterior region of the cell.

apokinetal A type of stomatogenesis in which the kinetosomes forming the ciliature of the oral region of the daughter cell have no apparent pre-association with either the parental somatic kineties or oral apparatus.

atrium A shallow depression in the oral region of certain hypostomes.

autogamy A sexual process in ciliates where self-fertilisation takes place. It differs from cytogamy in that no pairing takes place.

AZM *see* adoral zone of membranelles.

basket of trichites A group of rod-like elements used to support the cytopharynx (*syn.* cytopharyngeal apparatus).

bipolar kinetosomes Paired basal bodies of cilia.

brood pouch A cavity in which an endogenously produced embryo develops.

buccal cavity Typically a deep oral cavity or depression containing compound ciliary organelles such as membranelles and paroral membrane.

buccal overture The outer opening of a buccal cavity.

buccokinetal A type of stomatogenesis in which the kinetosomes forming the oral ciliature of the daughter cell have an apparent origin in the oral organelles of the parental buccal apparatus.

capitate Refers to tentacles bearing knobs at their distal ends.

caudal cilium A cilium (or group of cilia) longer than somatic cilia arising at the terminal end of the cell.

ciliary meridians Rows of cilia. More often than not refers to the rows of kinetosomes that are displayed by silver impregnation methods (*syn.* kinety).

cilium (*pl.* cilia) Fine hair-like organelles arranged over bodies of most ciliates. They beat in a coordinated fashion and are used for locomotion and feeding.

circumoral ciliature A line, arc or band of simple somatic cilia encircling all or part of the apical end (including cytostome) of the body.

cirrus (*pl.* cirri) A type of compound somatic ciliary structure consisting of a number of cilia which adhere together. The group of cilia forms a rod-like structure which usually narrows distally. They are typically used for locomotion.

clasper Posterior protoplasmic projection in a suctorian ciliate used to grip stalk of a peritrichous ciliate.

clavate Club-like, usually used to describe body shape when posterior end of body is distinctly narrower than the anterior. Also used to describe certain immobile cilia.

clone A population of organisms derived entirely from an individual cell.

conjugation A sexual process involving the exchange of genetic material during the pairing of two ciliates.

contractile vacuole A liquid-filled vacuole that serves to rid the cell of excess water.

cyrtos A usually curved cytopharyngeal apparatus supported by trichites. May be distinguished from the rhabdos type by its ultra-structure.

cytogamy A sexual process in ciliates where self-fertilisation takes place. Although two cells pair as in conjugation there is no exchange of genetic material between the sexes.

cytopharyngeal apparatus *Syn.* basket of trichites.

cytopharynx A non-ciliated tube leading from the cytostome into the cytoplasm. Sometimes supported by trichites.

cytoproct The generally permanent opening in the pellicle of the cell through which the undigested remains of food are voided to the exterior.

cytostome The true oral aperture of the cell which denotes the end of any ciliation.

discoid A term describing a three-dimensional shape which is spherical or oval in outline but flattened laterally.

dorsal brush A distinctive group of cilia arising from a few specialised kineties that are arranged on the anterior dorsal surface of a ciliate.

endogenous bud A bud produced asexually inside the cell within a brood pouch. The buds develop cilia before emergence from the brood pouch.

equatorial Refers to the central region (or equator) of the cell.

evaginative buds Buds that are produced internally but are evaginated before they gain cilia and are therefore released externally.

exogenous bud A bud that develops on the outside of the mother cell.

fascicle A bundle or group of several tentacles arising from a localised area on the cell.

food vacuole Intracellular vacuoles containing food within which the latter is digested. They form at the base of the cytopharynx and are discharged at the cytoproct.

haplokinety A double row of kinetosomes joined in a zig-zag fashion, generally with only the outermost row bearing cilia.

hypostomial frange A band of perioral ciliature characteristically found in certain hypostome ciliates. It varies from an extensive helical band of specialised postoral ciliature to a few pseudomembranelles lying close to the cytostome.

infraciliature The complete collection of somatic and oral kinetosomes and associated microfibrillar and microtubular structures that are revealed by silver impregnation methods.

infundibulum The inner part of the buccal cavity of certain ciliates, it may be funnel or tube shaped.

kinetodesmata Longitudinally orientated subpellicular cytoplasmic fibrils which arise close to the base of a kinetosome.

kinetosome Subpellicular basal body of a cilium.

kinety Single longitudinally orientated and functionally integrated row of somatic kinetosomes.

lappet A finger-like protoplasmic projection in the oral region.

lorica An external non-living housing or shell of a ciliate. They may be mucilaginous or pseudochitinous.

macroconjugant The larger of two conjugants. Often less mobile than the microconjugant and generally regarded as the female of the pair.

macronucleus The larger vegetative or trophic nucleus of a ciliate which controls the organism's phenotype. It is derived from the micronucleus during sexual reproduction.

membranelle   A compound ciliary organelle found on the left side of a buccal cavity or peristomial field.

metabolic   A term used to describe cells that are able to readily change their shape because of their plasticity.

microconjugant   The smaller of two conjugants. Often more mobile than the macroconjugant and generally regarded as the male of the pair.

micronucleus   The generative and smaller nucleus of ciliates. It is concerned with sexual processes.

moniliform   A term used to describe a macronucleus that takes the form of a string of beads or sausages.

mucocyst   Subpellicular sac or rod-shaped organelle which discharges mucus through pore in pellicle.

nematodesma (*pl.* **nematodesmata**)   Parallel bundles of microtubules (*syn.* trichite).

opisthe   Posterior daughter of a dividing ciliate cell.

parakinetal   A type of stomatogenesis in which the kinetosomes of the daughter cell are derived from one or more of the postoral kineties of the mother cell.

paroral membrane   A ciliary organelle lying along the right side or border of a buccal cavity.

peduncle   A non-contractile stalk.

pellicle   Outer living covering of a ciliate.

peniculus (*pl.* **peniculi**)   A band-like membranelle or compound ciliary organelle typically found travelling along the left wall of a buccal cavity.

perioral ciliature   Similar to circumoral ciliature but often used more loosely meaning any ciliature around the cytostome.

peristome   Usually used to mean the complete oral area (peristomial field) of peritrichs and certain spirotrichs where the buccal ciliature has been everted from a cavity and occupies much of the anterior portion of the body.

peristomial ciliature   Compound ciliary organelles in the region (usually surrounding) the peristome.

phoront   Stage in the life cycle of an apostome ciliate, where it is carried about on the carapace of a crustacean.

postoral frange   *Syn.* hypostomial frange.

preoral funnel   The expanded anterior funnel-like region of a chonotrich. It is ciliated internally and represents a vestibulum.

proter   Anterior daughter of a dividing ciliate.

pseudomembranelle   A loose term to describe oral or somatic ciliary complexes which superficially resemble membranelles but which on closer examination differ from them structurally.

pyriform   Pear shaped.

reniform   Kidney shaped.

rhabdos   A cytopharyngeal apparatus supported by straight trichites. May be distinguished from the cyrtos type by its fine structure.

rosette organelle   A rose-shaped septate structure of unknown function found near the cytostome in apostome ciliates.

rostrum   A general term referring to the apical end of a ciliate when it has a beak-like appearance. It is usually bent at an angle to the longitudinal body axis.

sapropelic   An aquatic habitat containing a high concentration of organic matter and a low concentration of dissolved oxygen (frequently anaerobic).

somatic ciliature   The cilia or compound ciliary organelles found anywhere on the body surface other than in the oral area.

stomatogenesis The process by which a new oral area is formed. The term is usually used with special reference to the formation of new oral ciliature.

stomatogenic kinety The parental kinety from which the oral ciliature of a daughter cell is formed.

subapical Area just below the apex of the body.

suture The linear space left between the ends of converging kineties.

swarmer Freely swimming, ciliated embryonic stage of a suctorian ciliate.

telokinetal A type of stomatogenesis in which the new oral ciliature is formed from all or some of the encircling somatic kinetosomes or from those comprising the kinetofragments.

tentacle Hollow tube-like extension of body of suctorian. They are extensible and retractable and are often knobbed (capitate). Tentacles are used for feeding.

terminal Extreme posterior region of the body.

tomite Small, free-swimming, non-feeding stage in the polymorphic life cycle of an apostome ciliate. It serves to spread the organism to a new 'host' and is the product of the multiple fission of an encysted trophont.

toxicyst Subpellicular slender tubular organelle found particularly in the carnivorous gymnostomes. Also found in the non-suctorial tentacles of certain other gymnostomes.

trichite A skeletal rod-like structure used to support the cytopharynx of some ciliates. Now often called a cytopharyngeal rod.

trichocyst A term used loosely throughout this volume to include mucocysts and trichocysts proper. The trichocysts *sensu stricto* are fibrous explosive organelles that are restricted to the peniculine hymenostomes and hypostome microthoracines.

trophont Feeding stage of an apostome ciliate. It feeds on the exuvial fluid of the moulted exoskelton of a crustacean.

undulating membrane *Syn.* the modern term paroral membrane.

vermiform Worm-like.

vestibulum A depression or cavity in the body containing cilia that are predominantly somatic in origin.

zoochlorellae Green mutualistic algae found in certain ciliates.

Acknowledgements

The author is indebted to the Natural Environment Research Council who provided a grant for artist fees, to Mr Harold Weston-Bird for his patient artistic talents in the preparation of the generic diagrams, to Professor John O. Corliss for his kind assistance particularly concerning some of the more obscure references, to Miss B. Grimes for the provision of information on the apostome ciliates and to my wife Polly who read the final manuscript.

References

Where two dates are given the first is the date of availability, the second (given in parenthesis) is that actually printed on the publication. All abbreviations follow those given in the *World List*.

André, E. 1914. Recherches sur la faune pélagique du Léman et description de nouveaux genres d'Infusoires. *Revue suisse Zool.* **22**(7), 179–93.

André, E. 1915. Contribution à l'étude de la faune infusorienne du Lac Majeur et description de formes nouvelles. *Revue suisse Zool.* **23**(4), 101–8.

Awerinzew, S. 1904. *Astrophrya arenaria* nov. gen., nov. spec. *Zool. Anz.* **27**, 425–6.

Bardele, C. F. 1970. Budding and metamorphosis in *Acineta tuberosa*. An electron microscopic study on morphogenesis in Suctoria. *J. Protozool.* **17**(1), 51–70.

Batisse, A. 1975a. Propositions pour une nouvelle systématique des Acinétiens (Ciliophora, Kinetofragminophora, Suctorida). *C.r. hebd. Séanc. Acad. Sci., Paris* **280**, 1797–800.

Batisse, A. 1975b. Propositions pour une nouvelle systématique des Acinétiens (Ciliophora, Kinetofragminophora, Suctorida). *C.r. hebd. Séanc. Acad. Sci., Paris* **280**, 2121–4.

Beers, C. D. and Sherwood, W. A. 1966. A new species of *Woodruffia*, a zoospore-ingesting ciliate occurring on the water mold *Saprolegnia*. *Trans. Am. microsc. Soc.* **85**, 528–36.

Bhatia, B. L. 1936. *The Fauna of British India. Including Ceylon and Burma. Protozoa: Ciliophora.* Taylor & Francis, London. 493pp.

Bick, H. 1972. *Ciliated Protozoa. An Illustrated Guide to the Species Used as Biological Indicators in Freshwater Biology.* World Health Organization, Geneva. 198pp.

Blochmann, F. 1886. Die Mikroskopische Thierwelt des Süsswassers. In *Die Mikroskopische Pflanzen – und Thierwelt des Süsswassers,* ed. O. Kirchner and F. Blochmann, Braunschweig, 1st edition, vol. 2. 122pp.

Blochmann, F. 1895. Die Mikroskopische Thierwelt des Süsswassers. In *Die Mikroskopische Pflanzen – und Thierwelt des Süsswassers,* ed. O. Kirchner and F. Blochmann, Braunschweig, 2nd edition, vol. 1. 94pp.

Bohatier, J. 1971 (1970). Structure et ultrastructure de *Lacrymaria olor* (O.F.M. 1786). *Protistologica* **6** (year 1970), 331–42.

Bohatier, J. and Detcheva, R. 1973. Observations sur la cytologie et sur l'ultrastructure du Cilié *Acropisthium mutabile* Perty, 1852. *C.r. Séanc. Soc. Biol.* **167**, 972–4.

Borror, A. 1972. Tidal marsh ciliates (Protozoa): morphology, ecology, systematics. *Acta Protozool.* **10**(2), 29–72.

Bory, de St Vincent J. B. 1826. *Essai d'une Classification des Animaux Microscopiques,* Paris. 104pp.

Bradbury, P. C. 1966. The morphology and life cycle of the apostomatous ciliate *Hyalophysa chattoni* n.g., n. sp. *J. Protozool.* **13**(2), 209–25.

371

Bradbury, P. C. and Clamp, J. C. 1973. *Hyalophysa lwoffi* sp. n. from the freshwater shrimp *Palaemonetes paludosus* and révision of the genus *Hyalophysa*. *J. Protozool.* **20**(2), 210–13.

Bramy, M. 1962. Un nouveau cilié trichostome *Kalometopia perronei* n.g., n. sp. *C.r. hebd. Séanc. Acad. Sci., Paris* **254**, 162–4.

Brodsky, A. 1925. Zwei neue holotriche Infusorien aus Turkestan. *Bull. Univ. Asie cent.* **8**, 40–4.

Buitkamp, U. 1975. Eine neubeschreibung von *Mycterothrix tuamotuensis* Balbiani, 1887 (Ciliophora, Colpodida). *Protistologica* **11**, 323–4.

Buitkamp, U. and Wilbert, N. 1974. Morphologie und Taxonomie einiger Ciliaten eines kanadischen Präriebodens. *Acta Protozool.* **13**, 201–10.

Burt, R. L. 1940. Specific analysis of the genus *Colpoda* with special reference to the standardization of experimental material. *Trans. Am. microsc. Soc.* **59**, 414–32.

Bütschli, O. 1887–89. Protozoa. Abt. III. Infusoria und System der Radiolaria. In *Klassen und Ordnung des Thiersreichs*, ed. H. G. Bronn and C. F. Winter, vol. 1, pp. 1098–2035. Leipzig.

Canella, M. F. 1951. Osservazioni morfologiche, biologiche e sistematiche su *Paradileptus estensis* sp. n. e su altri Tracheliidae (Holotricha). *Annali Univ. Ferrara* (N.S. Sez III) **1**(2), 81–170.

Canella, M. F. 1957. Studi i ricerche sui tentaculiferi nel quadro della biologia generale. *Annali Univ. Ferrara* (N. S. Sez III) **1**, 259–716.

Canella, M. F. 1960. Contributo ad una revisione dei generi *Amphileptus, Hemiophrys* e *Lionotus* (Ciliata, Holotricha, Gymnostomata). *Annali Univ. Ferrara* (N.S. Sez III) **2**(2), 47–95.

Certes, A. 1891. Note sur deux infusoires nouveaux des environs de Paris. *Mém. Soc. zool. Fr.* **4**, 536–41.

Chatton, E. and Beauchamp, P. de 1923. *Teuthophrys trisulca* n.g., n.sp., infusoire pélagique d'eau douce. *Archs Zool. exp. gén.* **61**, 123-9.

Chatton, E. and Lwoff, A. 1930. Imprégnation, par diffusion argentique, de l'infraciliature des ciliés marins et d'eau douce, après fixation cytologique et sans dessiccation. *C.r. Séanc. Soc. Biol.* **104**, 834–6.

Chatton, E. and Lwoff, A. 1935. Les ciliés apostomes. Morphologie, cytologie, éthologie, évolution, systématique. Première partie. Aperçu historique et général. Étude monographique des genres et des espèces. *Archs Zool. exp. gén.* **77**, 1–453.

Claff, C. L., Dewey, V. C. and Kidder, G. W. 1941. Feeding mechanisms and nutrition in three species of *Bresslaua*. *Biol. Bull. mar. biol. lab. Woods Hole* **81**, 221–34.

Claparède, E. and Lachmann, J. 1858 (1857). Études sur les infusoires et les rhizopodes. *Mém. Inst. natn. génev.* **5** (year 1857), 1–260.

Claparède, E. and Lachmann, J. 1859 (1858). Études sur les infusoires et les rhizopodes. *Mém. Inst. natn. génev.* **6** (year 1858), 261–482.

Claparède, E. and Lachmann, J. 1861 (1859–60). Études sur les infusoires et les rhizopodes. *Mém. Inst. natn. génev.* **7** (year 1859–60), 1–291.

Clément-Iftode, F. and Versavel, G. 1968 (1967). *Teutophrys trisulca* (Chatton, de Beauchamp) cilié planctonique rare. *Protistologica* **3** (year 1967), 457–64.

Cohn, F. 1866. Neue Infusorien im Seeaquarium. *Z. wiss. Zool.* **16**, 253–302.

Collin, B. 1906. Note préliminaire sur un Acinétien nouveau: *Dendrosomides paguri* n.g., n.sp. *Archs Zool. exp. gén.* (Sér. 4), **5** (Notes et Revue), LXIV–LXVII.

Collin, B. 1909. Diagnoses préliminaires d'acinétiens nouveaux ou mal connus. *C.r. hebd. Séanc. Acad. Sci., Paris* **149**, 1094–5.

Collin, B. 1911. Étude monographique sur les Acinétiens. I. Recherches expérimentales sur l'étendue des variations et les facteurs tératogènes. *Archs Zool. exp. gén.* **48,** 421–97.

Collin, B. 1912. Étude monographique sur les Acinétiens. II. Morphologie, Physiologie, Systématique. *Archs Zool. exp. gén.* **51,** 1–457.

Committee on Cultures, Society of Protozoologists, 1958. A catalogue of laboratory strains of free-living and parasitic protozoa. *J. Protozool.* **5**(1), 1–38.

Conn, H. W. 1905: A preliminary report on the protozoa of the fresh waters of Connecticut. *Bull. Conn. St. geol. nat. Hist. Surv.* **1**(2), 5–69.

Conn, H. W. and Edmondson, C. H. 1918. Flagellate and Ciliate protozoa (Mastigophora et Infusoria). In *Freshwater Biology,* ed. H. B. Ward and G. C. Whipple, pp. 238–300. Wiley & Sons, New York.

Corliss, J. O. 1953. Silver impregnation of ciliated protozoa by the Chatton-Lwoff technic. *Stain. Technol.* **28,** 97–100.

Corliss, J. O. 1958a. The systematic position of *Pseudomicrothorax dubius*, ciliate with a unique combination of anatomical features. *J. Protozool.* **5**(3), 184–93.

Corliss, J. O. 1958b. The phylogenetic significance of the genus *Pseudomicrothorax* in the evolution of holotrichous ciliates. *Acta biol. hung.* **8**(4), 367–88.

Corliss, J. O. 1960. The problem of homonyms among generic names of ciliated Protozoa, with proposals of several new names. *J. Protozool.* **7**(3), 269–78.

Corliss, J. O. 1975. Taxonomic characterization of the suprafamilial groups in a revision of recently proposed schemes of classification for the phylum Ciliophora. *Trans. Am. microsc. Soc.* **94**(2), 224–67.

Corliss, J. O. 1977. Annotated assignment of families and genera to the orders and classes currently comprising the Corlissian scheme of higher classification for the phylum Ciliophora. *Trans. Am. microsc. Soc.* **96,** 104–40.

Corliss, J. O. 1979. *The Ciliated Protozoa: Characterization, Classification and Guide to the Literature,* 2nd edition. Pergamon Press, Oxford. 455pp.

Cunha, A. M. da 1914. Contribuição para o cohecimento da fauna de Protozoarios do Brazil. *Mems Inst. Oswaldo Cruz* **6,** 169–79.

Czapik, A. 1971. Les observations sur *Platyophrya spumacola* Kahl. *Acta Protozool.* **8,** 363–6.

Deroux, G. 1965. Origine des cinéties antérieures, gauches et buccales dans le genre *Dysteria* Huxley. *C.r. hebd. Séanc. Acad. Sci., Paris* **260,** 6689–1.

Deroux, G. 1970. La série 'Chlamydonellienne' chez les Chlamydodontidae (Holotriches, Cyrtophorina Fauré-Fremiet). *Protistologica* **6**(2), 155–82.

Deroux, G. 1976. Le plan cortical des Cytophorida unité d'expression et marges de variabilité. II – Cytophorida a thigmotactisme ventral généralisé. *Protistologica* **12**(3), 483–500.

Deroux, G. 1977 (1976). Plan Cortical des Cyrtophorida III – Les structures différenciatrices chez les Dysteriina. *Protistologica* **12,** (year 1976), 505–38.

Deroux, G. 1978. The hypostome ciliate order Synhymeniida: from *Chilodontopsis* of Blochmann to *Nassulopsis* of Fauré-Fremiet. *Trans. Am. microsc. Soc.* **97**(4), 458–68.

Deroux, G., Iftode, F. and Fryd, G. 1974. Le genre *Nassulopsis* et les ciliés fondamentalement hypostomiens. *C.r. hebd. Séanc. Acad. Sci., Paris* **278,** 2153–6.

Detcheva, R. 1976. Particularités ultrastructurales du cilié *Cyrtolophosis mucicola* Stokes, 1885. *C.r. Séanc. Soc. Biol.* **170,** 112–14.

Dewey, V. and Kidder, G. W. 1940. Growth studies on ciliates VI. Diagnosis, sterilization and growth characteristics of *Perispira ovum*. *Biol. Bull. mar. biol. lab. Woods Hole* **79**, 255–71.

Dietz, G. 1964. Beitrag zur Kenntnis der Ciliatenfauna einiger Brackwasser-stumpel (Etangs) der Französischen Mittelmeerküste. *Vie Milieu* **15**, 47–93.

Diller, W. F. 1964. Observations on the morphology–life history of *Homalozoon vermiculare*. *Arch. Protistenk.* **107**, 351–62.

Dingfelder, J. H. 1962. Die Ciliaten vorübergerehender Gewässer. *Arch. Protistenk.* **105**, 509–658.

Dragesco, J. 1963. Révision du genre *Dileptus* Dujardin, 1841 (Ciliata Holotricha) (systématique, cytologie, biologie). *Bull. biol. Fr. Belg.* **97**(1), 103–45.

Dragesco, J. 1966a. Observations sur quelques ciliés libres. *Arch. Protistenk.* **109**, 155–206.

Dragesco, J. 1966b. Ciliés libres de Thonon et ses environs. *Protistologica* **2**(2), 59–95.

Dragesco, J. 1970. Ciliés libres du Cameroun. *AnnlsFac. Sci. Univ. féd. Cameroun* (Numéro Hors-série) **1970**, 1–141.

Dragesco, J. 1972. Ciliés libres de l'Ouganda. *AnnlsFac. Sci. Univ. féd. Cameroun* **9**, 86–126.

Dragesco, J., Blanc-Brude, R. and Skreb, Y. 1955. Morphologie et biologie d'un tentaculifère peu connu: *Heliophrya erhardi* (Reider) Matthes. *Bull. Microsc. appl.* (Sér. 2) **5**, 103–12.

Dragesco, J. and Dragesco-Kerneis, A. 1979. Ciliés muscicoles nouveaux ou peu connus. *Acta Protozool.* **18**(3), 401–16.

Dragesco, J., Fryd-Versavel, G., Iftode, F. and Didier, P. 1977. Le cilié *Platyophrya spumacola* Kahl, 1926: morphologie, stomatogenèse et ultrastructure. *Protistologica* **13**, 419–34.

Dragesco, J., Iftode, F. and Fryd-Versavel, G. 1974. Contribution à la connaissance de quelques ciliés holotriches rhabdophores; I. Prostomiens. *Protistologica* **10**(1), 59–75.

Dujardin, F. 1841a (1840). Mémoire sur une classification des infusoires en rapport avec leur organisation. *C.r. hebd. Séanc. Acad. Sci., Paris* **11**, 281–6.

Dujardin, F. 1841b. *Histoire Naturelle des Zoophytes. Infusoires*. De Fain et Thunot, Paris, 678pp.

Eberhard, E. 1862. Zweite Abhandlungen uber die Infusorienwelt. *Oster Programm der Realschule zu Coburg*, 1–34.

Edmondson, C. H. 1920. Protozoa of Devil's Lake Complex, North Dakota. *Trans. Am. microsc. Soc.* **39**(3), 167–98.

Ehrenberg, C. G. 1830 (1832). Beiträge zur Kenntnis der Organisation der Infusorien und ihrer geographischen Verbreitung, besonders in Siberien. *Abh. Akad. Wiss. DDR*, year 1832, 1–88.

Ehrenberg, C. G. 1833 (1835). Dritter Beiträge zur Erkenntnis grosser Organisation in der Richtung des Kleinsten Raumes. *Abh. Akad. Wiss. DDR*, year 1835, 145–336.

Ehrenberg, C. G. 1835 (1837). Zusätze zur Erkenntnis grossen organischer Ausbildung in den kleinsten thierischen Organismen. *Abh. Akad. Wiss. DDR*, year 1837, 151–80.

Ehrenberg, C. G. 1838. *Die Infusionsthierchen als Vollkommene Organismen*. Leipzig. 612pp.

Engelmann, T. W. 1862. Zur Naturgeschichte der Infusionsthiere: Beiträge zur Entwicklungsgeschichte der Infusorien. *Z. wiss. Zool.* **11**, 347–93.

Englemann, T. W. 1876. Ueber Entwicklung und Fortpflanzung der Infusorien. *Morph. Jb.* **1**, 573–635.

Fabre-Domergue, P. L. 1888. Recherches anatomiques et physiologiques sur les Infusoires Ciliés. *Annls Sci. nat. Zool.* (Sér. 7), **5**, 1–140.

Farkas, B. 1924. Beiträge zur Kenntnis der Suctorien. *Arch. Protistenk.* **48**, 125–35.

Fauré-Fremiet, E. 1908. Sur deux infusoires nouveaux de la famille des Trachelidae. *Bull. Soc. zool. Fr.* **33**, 13–16.

Fauré-Fremiet, E. 1924. Contribution à la connaissance des infusoires planktoniques. *Bull. biol. Fr. Belg.* Suppl. **6**, 1–171.

Fauré-Fremiet, E. 1944a (1943). Commensalisme et adaptation chez un acinétien: *Erastophrya chattoni*, n.gen., n.sp. *Bull. Soc. zool. Fr.* **68**, (year 1943), 145–7.

Fauré-Fremiet, E. 1944b. Polymorphisme de l'*Enchelys mutans* (Mermod). *Bull. Soc. zool. Fr.* **69**, 212–19.

Fauré-Fremiet, E. 1945. Polymorphisme du *Monodinium vorax* nov. sp. *Bull. Soc. zool. Fr.* **70**, 69–79.

Fauré-Fremiet, E. 1950. Méchanisme de la morphogenèse chez quelques ciliés gymnostomes hypostomiens. *Archs Anat. microsc. Morph. exp.* **39**, 1–14.

Fauré-Fremiet, E. 1959. La famille des Nassulidae (Ciliata, Gymnostomatida) et le genre *Nassulopsis* n. gen. *C.r. hebd. Séanc. Acad. Sci., Paris* **249**, 1429–33.

Fauré-Fremiet, E. 1965. Morphologie des Dysteriidae (Ciliata, Cyrtophorina). *C.r. hebd. Séanc. Acad. Sci., Paris* **260**, 6679–84.

Fauré-Fremiet, E. 1967a. La frange ciliaire des Nassulidae (Ciliata, Cyrtophorina) et ses possibilities évolutives. *C.r. hebd. Séanc. Acad. Sci., Paris* **264**, 68–72.

Fauré-Fremiet, E. 1967b. Le genre *Cyclogramma* Perty, 1852, *J. Protozool.* **14**(3), 456–64.

Fauré-Fremiet, E. and André, J. 1965a. Étude au microscopique de *Tillina praestans* Penard (Cilié Trichostomatida). *Archs Zool. exp. gén.* **105**, 345–53.

Fauré-Fremiet, E. and André, J. 1965b. L'organisation du cilie gymnostome *Plagiocampa ovata*, Gelei. *Archs Zool. exp. gén.* **105**, 360–7.

Fauré-Fremiet, E. and Ganier, M. C. 1970 (1969). Morphologie et structure fine du cilié *Chaenea vorax* Quenn. *Protistologica* **5**(3), 353–61.

Fauré-Fremiet, E. and Hamard, M. 1940. Composition chimique du tégument chez *Coleps hirtus* Nitzsch. *Bull. biol. Fr. Belg.* **78**, 136–42.

Fernandez-Galiano, D. 1976. Silver impregnation of ciliated protozoa: procedure yielding good results with the pyridinated carbonate method. *Trans. Am. microsc. Soc.* **95**, 557–60.

Fernandez-Galiano, D. 1978. La position systématique de *Bursaria truncatella*. Un nouveau ordre des Ciliés. *J. Protozool.* **25**(3), 54A.

Foissner, W. 1972. Das Silberliniensystem von *Placus luciae* (Kahl, 1926) (Ciliata, Enchelyidae). *Arch. Protistenk.* **114**, 83–95.

Foissner, W. 1978a. Das Silberliniensystem und die Infraciliatur der Gattungen *Platyophrya* Kahl, 1926, *Cyrtolophosis* Stokes, 1885 und *Colpoda* O.F.M., 1786: Ein Beiträg zur Systematik der Colpodid (Ciliata, Vestibulifera). *Acta Protozool.* **17**, 215–31.

Foissner, W. 1978b. Morphologie, Infraciliature und Silberliniensystem von *Plagiocampa rouxi* Kahl, 1926 (Prostomatida. Plagiocampidae) und *Balanonema sapropelica* nov. spec. (Philasterina, Loxocephalidae). *Protistologica* **14**, 381–9.

Foissner, W. 1979. Ökologische und systematische Studien über das Neuston alpiner kleingewässer, mit besonderer Berücksichtigung der Ciliaten. *Int. Revue ges. Hydrobiol.* **64**(1), 99–140.

Foissner, W. 1980 (1979). Morphologie, Infraciliatur und Silberliniensystem von

Phascolodon vorticella Stein, *Chlamydonella alpestris* nov. spec. und *Trochilia minuta* (Roux) (Ciliophora, Cytophorida). *Protistologica* **15**(4), 557–563.

Frenzel, J. 1891. Uber einige merkwürdige Protozoen Argentiniens. *Z. wiss. Zool.* **53**, 334–60.

Fresenius, G. 1858. Beträge zur Kenntnis mikroscopischer Organismen. *Abh. senckenb. naturforsch. Ges.* **2**(2), 211–42.

Fryd-Versavel, G., Iftode, F. and Dragesco, J. 1976 (1975). Contribution à la connaissance de quelques ciliés gymnostomes. II Prostomiens, Pleurostomiens: morphologie, stomatogenèse. *Protistologica*. **11**(4), 509–30.

Gajewskaja, N. 1928. Sur quelques infusoires pélagiques nouveaux du lac Baïkal. *Dokl. Akad. Nauk SSR* **1928A**(23), 476–8.

Gajewskaja, N. 1929. Über einige seltene Infusorien aus dem Baikalsee. *Izv. Akad. Nauk SSSR* **9**(series 7), 845–54.

Gajewskaja, N. 1933. Zur Oekologie, Morphologie und Systematik, der Infusorien des Baikalsees. *Zoologica, Stuttg.* **32**(83), 1–298.

Gelei, J. 1933. Beiträge zur Ciliatenfauna der Umgebung von Szeged II. Vier *Bryophyllum*-Arten. *Arch. Protistenk.* **81**, 201–30.

Gelei, J. 1939. Beiträge zur Ciliatenfauna der Umgebung von Szeged. X. *Nassula heterovesiculata* n.sp. *Acta biol., Szeged* **5**, 92–8.

Gelei, J. 1950. Die Marynidae der Sodagewässer in der Nähe von Szeged. XIV. Beiträg zur Ciliatenfauna Ungarns. *Hidrol. Kozl.* **30**, 107–19; 157–8.

Gellért, J. 1950a. A *Cirrophrya haptica* n. gen. n. sp. alkata és élettana. *Annls. biol. Univ. szeged* **1**, 295–312.

Gellért, J. 1950b. Neue Colpoda-Art unter der Flechte von Felsen. *Annls biol. Univ. szeged* **1**, 313–19.

Gellért, J. 1955. Die Ciliaten des sich unter der Flechte 'Parmalia saxatilis Mass' gebildeten Humus. *Acta Biol. Hung.* **6**, 77–111.

Gerassimova, Z. P., Sergejeva, G. I. and Seravin, L. N. 1979. Ciliary and fibrillar structures of the ciliate *Bursaria truncatella* and its systematic postition. *Acta Protozool.* **18**(3), 355–70.

Ghosh, E. 1928. Two new ciliates from sewer water. *Jl R. microsc. Soc.* **48**(4), 382–4.

Gieman, Q. M. 1931. Morphological variations in *Coleps octospinus* (Protozoa, Ciliata) *Trans. Am. microsc.* **50**, 136–43.

Girgla, H. 1971. Cortical anatomy and morphogenesis in *Homalozoon vermiculare* (Stokes). *Acta Protozoologica* **8**, 355–61.

Gönnert, R. 1935. Über Systematik, Morphologie, Entwicklungsgeschichte und Parasiten einiger Dendrosomidae nebst Beschreibung zweier neuer Suktorien. *Arch. Protistenk.* **86**, 113–54.

Grain, J. 1970. Structure et ultrastructure de *Lagynophrya fusidens* Kahl, 1927. *Protistologica* **6**(1), 37–51.

Grain, J. and Golinska, K. 1970 (1969). Structure et ultrastructure de *Dileptus cygnus* Claparède et Lachmann, 1859, cilié holotriche gymnostome. *Protistologica* **5**(2), 269–90.

Grain, J., Iftode, F. and Fryd-Versavel, G. 1980 (1979). Étude des infraciliatures somatique et buccale de *Bryophrya bavariensis* et considérations systématiques. *Protistologica* **15**(4), 581–95.

Grain, J., Peck, R. K., Didier, P. and Rodrigues de Santa Rosa, M. 1976. Importance de la microscopie électronique dans les études de systématique chez les Unicellulaires; un example: les Ciliés du genre *Cyclogramma*. *C.r. hebd. Séanc. Acad. Sci., Paris* **282**, 735–8.

Grandori, R. and Grandori, L. 1934. Studî sui Protozoi del terreno. *Boll. Lab. Zool. agr. Bachic. R. Ist. sup. agr. Milano* **5**(1934), 1–339.

Grandori, R. and Grandori, L. 1935 (1934). Nuovi Protozoi del terreno agrario (Nota preliminare). *Boll. Lab. Zool. agr. Bachic. R. Ist. sup. agr. Milano* **4** (1934), 64–80.

Grimes, B. 1976. Notes on the distribution of *Hyalophysa* and *Gymnodinioides* on crustacean hosts in coastal North Carolina and a description of *Hyalophysa trageri* sp.n. *J. Protozool.* **23**(2), 246–51.

Grolière, C. A. 1975. La stomatogenèse du cilié *Platyophrya spumacola* Kahl, 1927, son intérêt pour la compréhension de la diversification buisonnante des Kinetofragminophora de Puytorac *et coll. C.r. hebd. Séanc. Acad. Sci., Paris* **280**, 861–4.

Grolière, C. A. 1977 (1975–76). Contribution à l'étude des ciliés des sphaignes et des étendues d'eau acides I. Description de quelques espèces de gymnostomes, hypostomes, hymenostomes et heterotriches. *Annls Stn biol. Besse-en-Chandesse* **10** (1975–76), 265–97.

Gruber, A. 1879. Vorläufige Mitthelung über neue Infusorien. *Zool. Anz.* **2**, 518–19.

Gruber, A. 1880. Neue Infusorien. *Z. wiss. Zool.* **33**, 439–66.

Gruber, A. 1884. Die Protozoen des Hafens von Genua. *Nova Acta Acad. Caesar. Leop. Carol.* **46**, 473–539.

Guilcher, Y. 1951. Contribution à l'étude des ciliés gemmipares, chonotriches et tentaculifères. *Annls Sci. nat. Zool.* (Ser. 11), **13**, 33–132.

Hamburger, C. 1903. Beiträge zur Kenntnis von *Trachelius ovum. Arch. Protistenk.* **2**, 445–74.

Hartog, M. 1902. Notes on Suctoria. *Arch. Protistenk.* **1**, 372–4.

Heuss, K. and Wilbert, N. 1973. Zur Morphologie und Ökologie von *Trochilia minuta* Roux, 1901 (Ciliata, Cyrtophorina). *Gewäss. Abwäss.* **52**, 32–43.

Hickson, S. J. and Wadsworth, J. T. 1909. *Dendrosoma radians,* Ehrenberg. *Q.Jl microsc. Sci.* **54**, 141–83.

Hitchen, E. T. and Butler, R. D. 1972. A redescription of *Rhyncheta cyclopum* Zenker (Ciliatea, Suctorida). *J. Protozool.* **19**(4), 597–601.

Hovasse, R. 1938. Quelques particularités de la biologie et de la morphologie d'un *Lacrymaria olor* (O. F. Müller) marin de Roscoff. *Trav. Stn zool. Wimereux* **13**, 359–65.

Hukui, T. 1956. Taxonomical, morphological and ecological studies on *Colpoda saprophila. Bull. Fukuoka Univ. Educ.* **6**, 81–101.

Hull, R. W. 1954. The morphology and life cycle of *Solenophrya micraster* Penard, 1914. *J. Protozool.* **1**(2), 93–104.

Huxley, T. H. 1857. On *Dysteria,* a new genus of infusoria. *Quart. J. micr. Sci.* **5**, 78–82.

Jankowski, A. W. 1964. Morphology and evolution of Ciliophora III. Diagnoses and phylogenesis of 53 sapropelebionts, mainly of the order Heterotrichida. *Arch. Protistenk.* **107**, 185–294.

Jankowski, A. W. 1967a. New Genera of tentacled Infusoria (Suctoria). *Mater. V Konf. uch. Moldavii, Zool.* pp. 35–6.

Jankowski, A. W. 1967b. Taxonomy of the genus *Chilodonella* and a new proposed genus *Trithigmostoma* gen. nov. *Zool. Zh.* **46**(8), 1247–50.

Jankowski, A. W. 1967c. New genera and subgenera of classes Gymnostomea and Ciliostomea. *Mater. V. Konf. uch. Moldavii, Zool.* p. 36.

Jankowski, A. W. 1968. Taxonomy of the suborder Nassulina Jank., 1967 (Ciliophora, Ambihymenida). *Zool. Zh.* **47**(7), 990–1001.

Jankowski, A. W. 1973a. Taxonomic sketch of class Suctoria Claparéde and Lachmann, 1858. *Akad. Nauk SSSR Zool. Inst.* March **1973**, 30–1.

Jankowski, A. W. 1973b. *Fauna of the USSR. Infusoria Subclass Chonotricha. Akad. Nauk SSSR,* Nauka, Leningrad **2**(1), 355pp.

Jankowski, A. W. 1973c. Free-living Ciliophora. 1. *Myriokaryon* gen. n., giant planktonic holotrich. *Zool. Zh.* **52**(3), 424–8.

Jankowski, A. W. 1975. A conspectus of the new system of subphyllum Ciliophora Doflein, 1901. In *Account of Scientific Sessions on Results of Scientific Work Year 1974: Abstracts of Reports,* (ed. U.S. Balashov), pp. 26–7, *Tez. Dokl. zool. Inst. Akad. Nauk SSSR.*

Johnson, W. H. and Larson, E. 1938. Studies on the morphology and life history of *Woodruffia metabolica,* nov. sp. *Arch. Protistenk.* **90**, 383–92.

Jordan, A. 1974. Morphologie et biologie de *Prorodon discolor* Ehr.-Blochm.-Schew. *Acta Protozool.* **13**(2), 5–8.

Kahl, A. 1926. Neue und wenig bekannte Formen der holotrichen und heterotrichen Ciliaten. *Arch. Protistenk.* **55**, 197–438.

Kahl, A. 1927. Neue und ergänzende Beobachtungen holotricher Ciliaten. *Arch. Protistenk.* **60**, 34–129.

Kahl, A. 1930. Neue und ergänzende Beobachtungen holotricher Infusorien II. *Arch. Protistenk.* **70**, 313–416.

Kahl, A. 1930–35. Urtiere oder Protozoa. I: Wimpertiere oder Ciliata (Infusoria), eine Bearbeitung der freilebenden und ectocommensalen Infusorien der Erde, unter Ausschluss der marinen Tintinnidae. In *Die Tierwelt Deutschlands,* ed. F. Dahl, Teil **18** (year 1930), **21** (1931), **25** (1932), **30** (1935), 886 pp. G. Fischer, Jena.

Kahl, A. 1934. Suctoria. In *Die Tierwelt der Nord- und Ostsee,* ed. G. Grimpe and E. Wagler, Leif **26** (Teil II, C$_5$), pp. 184–226. Leipzig.

Kaneda, M. 1953. *Chlamydodon pedarius* n. sp. *J. Sci. Hiroshima Univ.* **14**, 51–5.

Kaneda, M. 1960. Phase contrast microscopy of cytoplasmic organelles in the gymnostome ciliate *Chlamydodon pedarius. J. Protozool* **7**(4), 306–13.

Kent, W. S. 1880–82. *A Manual of the Infusoria,* vols. I–III, David Bogue, London. 913pp.

Kink, J. 1972. Observations on the morphology and oral morphogenesis during regeneration of ciliate *Lacrymaria olor* (O.F.M. 1786) (Holotricha, Gymnostomatida). *Acta Protozool.* **10**(9), 205–13.

Kirby, H. 1950. *Materials and Methods in the Study of Protozoa.* University of California Press, Berkley and Los Angeles; Cambridge University Press. 72pp.

Klein, B. M. 1930. Das Silberliniensystem der Ciliaten. Weitere Ergebniss IV. *Arch. Protistenk.* **69**, 235–326.

Klein, B. M. 1958. The 'dry' silver method and its proper use. *J. Protozool.* **5**(2), 99–103.

Kormos, J. 1935. A *Prodiscophrya collini* (Root) ivari kétalakúsága és conjugatiója. *Allat. Közl.* **32**, 152–68.

Kormos, J. 1938. Fejlődéstani vizsgálatok a szivókásokon (Suctoria). *Mat. természettud. Közl.* **38**(1), 1–95.

Kormos, J. 1958. Phylogenetische Untersuchungen an Suctorien. *Acta biol. hung.* **9**, 9–23.

Kormos, J. 1960. Remarks on members of the suctorian family Discophryidae. I. '*Catharina' florea. J. Protozool.* **7**(suppl.), 21–2.

Kormos, J. and Kormos, K. 1956a. New investigations on the sexual dimorphism of the *Prodiscophrya. Acta biol. hung.* **7**, 109–25.

Kormos, J. and Kormos, K. 1956b. Determination in der Entwicklung der Suctorien I. Die Determination der Stelle der Embroorganisierung. *Acta biol. hung.* **7**, 365–83.

Kormos, J. and Kormos, K. 1956c. Die Ontogenes der Protozoen. *Acta biol. hung.* **7**, 385–402.

Kormos, J. and Kormos, K. 1957. Die Entwicklungsgeschlichtliche Grundlagen des Systems der Suctorien I. *Acta zool. hung.* **3**, 147–62.

Kormos, J. and Kormos, K. 1958a. Determination in der Entwicklung der Suctorien II. Neue untersuchungen über determinativen zusammenhang zwischen der schwärmerbildung und der metamorphose. *Acta biol. hung.* **9**, 25–45.

Kormos, J. and Kormos, K. 1958b. Uber die *Pseudogemma*-Frage. *Acta zool. hung.* **4**, 157–66.

Kormos, J. and Kormos, K. 1960a. Die Bedeuntung des Stadiums des Schwärmers der Suctoria in der Phylogense und Systematik. *Allatt. Közl.* **47**, 97–104.

Kormos, J. and Kormos, K. 1960b. Direkt Beobachtungen der Kernveränderungen der Konjugation von *Cyclophrya katherinae* (Ciliata: Protozoa). *Acta biol. hung.* **10**, 373–94.

Kormos, J. and Kormos, K. 1961. Phylogenetische Wertung Konvergenter und Divergenter Eigenschaften bei den Suctorien. *Acta biol. hung.* **11**, 335–8.

Kormos, K. 1958. Die Biologie von *Urnula*. *Urnula turpissima* n. sp. *Acta zool. hung.* **4**, 167–89.

Kudo, R. R. 1966. *Protozoology*, 5th edition. Charles C. Thomas, Springfield, Illinois. 1174pp.

Lachmann, J. 1859. Sitzungsberichte vom 6 Juli 1859. *Verh. naturh. Ver. preuss. Rheinl.* **16**, 91–4.

Lackey, J. B. 1925. Studies on the biology of sewage disposal. The fauna of Imhoff tanks. *Bull. N.J. agric. Exp. Sta.* no **417**, 1–39.

Lanners, H. N. 1973. Beobachtungen zur Konjugation von *Heliophrya* (*Cyclophrya*) *erhardi* (Rieder) Matthes, (Ciliata, Suctoria). *Arch. Protistenk.* **115**, 370–85.

Lauterborn, R. 1894. Ueber die Winterfauna einiger Gewässer der Oberrheinebene. Mit Beschreibungen neuer Protozën. *Biol. Zbl.* **14**, 390–8.

Lauterborn, R. 1898. Zwei neue Protozoen aus dem Gebeit des Oberrheins. *Zool. Anz.* **21**, 145–9.

Lauterborn, R. 1901. Die 'sapropelische' Lebewelt. *Zool. Anz.* **24**, 50–5.

Levander, K. M. 1900. Zur Kenntnis des Lebens in den stehenden Kleingewässern auf den Skäreninseln. *Acta Soc. Fauna Flora fenn.* **18**(6), 1–107.

López-Ochoterena, E. 1964. Mexican ciliated protozoa III. *Hypophrya fasciculata* gen. nov., sp. nov. (Ciliata: Suctorida). *J. Protozool.* **11**(2), 222–4.

Lynn, D. H. 1976. Comparative ultrastructure and systematics of the Colpodida. Fine structural specialisations associated with large body size in *Tillina magna* Gruber 1880. *Protistologica* **12**(4), 629–48.

McCoy, J. W. 1974. Biology and systematics of the ciliate genus *Cyrtolophosis* Stokes, 1885. *Acta Protozool.* **13**, 41–51.

Mackinnon, D. L. and Hawes, R. S. J. 1961. *An Introduction to the Study of Protozoa*. Clarendon Press, Oxford. 506pp.

Maskell, W. M. 1886. On the freshwater Infusoria of the Wellington District. *Trans. Proc. N.Z. Inst.* **19**, 49–61.

Maskell, W. M. 1887. On the freshwater Infusoria of the Wellington District. *Trans. Proc. N.Z. Inst.* **20**, 4–19.

Matthes, D. 1950. Beitrag zur Peritrichenfauna der Umgebung Erlangens. *Zool. Jb. (Syst)* **78**, 573–640.

Matthes, D. 1953. Suktorienstudien V. Die Zwischen Obligat Gebundenen Discophryen und Ihren Trägern Bestehenden Beziehungen. *Z. Morph. Okol. Tiere* **42**, 307–32.

Matthes, D. 1954a. Suktorienstudien I. Beitrag zur Kenntnis der Gattung *Discophrya* Lachmann. *Arch. Protistenk.* **99**, 187–226.

Matthes, D. 1954b. Suktorienstudien II. Uber obligatorisch symphorionte *Discophrya*-Arten. *Zool. Anz.* **152**, 106–21.

Matthes, D. 1954c. Suktorienstudien III. *Discophrya lichtensteinii* (Claparède u. Lachmann) Collin 1912, ein au Halipliden und Dytisciden gebundens Suktor. *Zool. Anz.* **152**, 252–62.

Matthes, D. 1954d. Suktorienstudien IV. Neue obligatorisch symphoriont mit Wasserkäfern vergesellschaftete *Discophrya*-Arten. *Zool. Anz.* **153**, 76–88.

Matthes, D. 1954e. Suktorienstudien VI. Die Gattung *Heliophrya* Saedeleer and Tellier 1929. *Arch. Protistenk.* **100**, 143–52.

Matthes, D. 1954f. *Discophrya buckei* (Kent), eine formenreiche Art der linguifera-Gruppe. *Zool. Anz.* **153**, 242–50.

Matthes, D. 1956. Suktorienstudien VIII. *Thecacineta calix* (Schröder 1907) (Thecacinetidae nov. fam.) und ihre Fortpflanzung durch Vermoid-Swärmer. *Arch. Protistenk.* **101**, 477–528.

Matthes, D. and Plachter, H. 1978. Das Sauginfusor *Spelaeophrya polypoides* (Daday) (Ciliata, Suctoria). *Arch. Protistenk.* **120**, 190–205.

Matthes, D. and Stiebler, G. 1970. Susswassersuktorien auf Arachniden. *Arch. Protistenk.* **112**, 65–70.

Mermod, G. 1914. Recherches sur la fauna infusorienne des tourbières et des eaux voisines de Sainte-Croix (Jura vaudois). *Revue suisse Zool.* **22**(3), 31–114.

Minkiewicz, R. 1912. Ciliata chromatophora, nouvel ordre d'infusoires à morphologie et reproduction bizarres. *C.r. hebd. Séanc. Acad. Sci., Paris* **155**, 513–15.

Miyashita, Y. 1933. Studies on a freshwater Foettingeriid ciliate. *Hyalospira caridinae* n.g. n.sp. *Jap. J. zool.* **4**, 439–60.

Müller, O. F. 1773. *Vermium Terrestrium et Fluviatilium, seu Animalium Infusorium, Helminthicorum et Testaceorum, non Marinorum, Succincta Historia.* Havniae et Lipsiae. 135pp.

Nitzsch, C. L. 1827. In *Allgemeine Encyclopaedie,* ed. J. C. Ersch and J. G. Gruber, vol. **16**, p. 69.

Njiné, T. 1979a. Étude ultrastructurale du cilié *Kuklikophrya dragescoi* gen. n., sp. n. *J. Protozool.* **26**(4), 589–98.

Njiné, T. 1979b. Compléments à l'étude des ciliés libres du Cameroun. *Protistologica* **15**(3), 343–54.

Njiné, T. 1980 (1979). Structure et morphogenèse buccales du cilié *Leptopharynx* (Mermod, 1914). *Protistologica* **15**(4), 459–65.

Noland, E. L. 1925. A review of the genus *Coleps,* with descriptions of two new species. *Trans. Am. microsc. Soc.* **44**(1), 3–13.

Nozawa, K. 1938. Some new fresh-water Suctoria. *Annotnes zool. jap.* **17**, 247–59.

Nozawa, K. 1939. Two new species of the freshwater suctorians, *Pottsia* and *Metacineta. Annotnes zool. jap.* **18**, 58–64.

Pätsch. B. 1974. Die Aufwuchsciliaten des Naturalehrparks hans Wildenrath. Monographische Bearbeitung der Morphologie und Ökologie. *Arb. Inst. landw. Zool. Bienenkde* **1**, 1–78.

Patterson, D. J. 1978. *Kahl's Keys to the Ciliates* (A translation of the keys to the level of subgenus, see Kahl, 1930–35). University of Bristol. 166pp.

Peck, R. K. 1975 (1974). Morphology and morphogenesis of *Pseudomicrothorax,* *Glaucoma* and *Dexiotricha,* with emphasis on the types of stomatogenesis in holotrichous ciliates. *Protistologica* **10**(3), 333–69.

Penard, E. 1914. Un curieux Infusoire. *Legendrea bellerophon. Revue suisse Zool.* **22**(13), 407–32.

Penard, E. 1917. Le genre *Loxodes. Revue suisse Zool.* **25**(16), 453–89.

Penard, E. 1920. Études sur les infusoires tentaculifères. *Mèm. Soc. Phys. Hist. nat. Genève* **39**, 131–227.

Penard, E. 1922. *Études sur les infusoires d'eau douce.* Georg et Cie, Gènève. 331pp. (The copy in the British Museum (Natural History) was accessed in 1921, it is not known how many other prepublication copies were available.)

Pérez, C. 1903. Sur un acinétien nouveau, *Lernaeophrya capitata,* trouvé sur le *Cordylophora lacustris. C.r. Séanc. Soc. Biol.* **55**, 98–100.

Perty, M. 1852. *Zur Kenntnis Kleinster Lebensformen nach Bau, Funktionem, Systematik mit Specialverzeichneiss der in der Schweiz beobacteten.* Jent u. Reinert, Bern. 228pp.

Prelle, A. 1961. Contribution à l'étude de *Leptopharynx costatus* (Mermod) (Infusoire Cilié), *Bull. biol. Fr. Belg.* **95**, 731–52.

Prelle, A. 1962. Position systématique du cilié holotriche *Leptopharynx costatus* Mermod, 1914. *C.r. hebd. Séanc. Acad. Sci., Paris* **254**, 4071–3.

Prelle, A. 1963. Bipartition et morphogenèse chez le Cilié Holotriche *Woodruffia metabolica* Johnson et Larson, 1938. *C.r. hebd. Séanc. Acad. Sci., Paris* **257**, 2148–51.

Prelle, A. 1968. Ultrastructures corticales du cilié holotriche *Drepanomonas dentata* Fresenius, 1858. *J. Protozool.* **15**(3), 517–20.

Puytorac, P. de 1965. Sur l'ultrastructure du complexe cytostome-cytopharynx chez le Cilié Holotriche *Holophrya vesiculosa* Kahl. *C.r. Séanc. Soc. Biol.* **159**(3), 661–3.

Puytorac, P. de and Njiné, T. 1971 (1970), Sur l'ultrastructure des *Loxodes* (Ciliés Holotriches). *Protistologica* **6**(4), 427–44.

Puytorac, P. de, Perez-Paniagua, F. and Perez-Silva, J. 1979. A propos d'observations sur la stomatogenèse et l'ultrastructure du cilié *Woodruffia metabolica* (Johnson et Larson, 1938). *Protistologica* **15**(2), 231–43.

Puytorac, P. de and Rodrigues de Santa Rosa, M. 1976 (1975). Observations cytologiques sur le cilié gymnostome *Loxophyllum meleagris* Duj., 1841. *Protistologica* **11**(3), 379–90.

Puytorac, P. de and Savoie, A. 1968. Observations cytologiques et biologiques sur *Prorodon palustris* nov. sp. *Protistologica* **4**(1), 53–60.

Quennerstedt, A. 1867. Bidrag till Sveriges Infusorie-fauna. *Acta Univ. lund.* **4**(7), 1–48.

Riordan, G. P. and Small, E. B. 1975. Stomatogenesis of the gymnostome ciliate, *Placus buddenbrocki. J. Protozool.* **22**(3), 17A.

Robin, C. 1879. Mémoire sur la structure et la reproduction de quelques infusoires tentaculés, suceurs et flagellés. *J. Anat. Physiol., Paris* **15**, 529–83.

Rodrigues de Santa Rosa, M. and Didier, P. 1976 (1975). Remarques sur l'ultrastructure du cilié gymnostome *Monodinium balbiani* (Fabre-Domergue, 1888). *Protistologica* **11**(4), 469–79.

Root, F. M. 1914. Reproduction and reactions to food in the suctorian *Podophrya Collini* n.sp. *Arch. protistenk.* **35**, 164–96.

Roux, J. 1899. Observations sur quelques infusoires ciliés des environs de Genève avec la description de nouvelles espèces. *Revue suisse Zool.* **6**, 557–636.

Roux, J. 1901. *Faune Infusorienne des Eaux Stagnantes des Environs de Genève.* Kündig, Genève. 148pp.

Saedeleer, H. de and Tellier, L. 1930 (1929). *Heliophrya Collini* n.g., n.sp., acinétien d'eau douce. *Annis Soc. r. zool. belg.* **60**, 12–15.

Sand, R. 1895, Les Acinétiens. *Annls Soc. belge Microsc.* **19**, 121–87.

Sand, R. 1899–1901. Étude monographique sur le groupe des Infusoires Tentaculifères. *Annls Soc. Belge Microsc.* **24**, 59–189; **25**, 1–204; **26**, 13–119.

Savi, L. 1913. Nuovi Ciliofori Appartenenti alla microfauna del lago-stagno craterico di Astroni. *Monitore zool. ital.* **24**, 95–100.

Savoie, A. 1957. Le cilié *Trichopelma agilis* n.sp. *J. Protozool.* **4**(4), 276–80.

Schewiakoff, W. 1892. Ueber die Geographische Verbreitung der Süsswasser-Protozoën. *Verh. naturh. -med. Ver. Heidelb.* (NS) **4**(5), 544–67.

Schewiakoff, W. 1893. Ueber die Geographische Verbreitung der Süsswasser-Protozoën. *Zap. Akad. Nauk SSSR* (Sér. 7) **41**, 1–201.

Schmidt, W. J. 1920. *Sphaerobactrum warduae,* ein kettenbildender Ciliat. *Arch. Protistenk.* **40**, 230–52.

Schmidt, W. J. 1921. Untersuchungen über Bau und Lebenserscheinungen von *Bursella spumosa,* einem neuen Ciliaten. *Arch. Mikrosk. Anat.* **95**(1), 1–36.

Schrank, F. Von P. 1803. *Fauna boica.* **3**(2), 1–372.

Shulz, E. 1931. Beiträge zur Kenntnis mariner Suctorien III. *Zool. Anz.* **97**, 289–92.

Skreb-Guilcher, Y. 1955. Quelques remarques sur un protozoaire cilié, le tentaculifére *Stylocometes digitatus* Stein. *Bull. Microsc. appl.* (Sér. 2) **5**, 118–21.

Smith, J. C. 1899. Notices of some undescribed Infusoria from the infusorial fauna of Louisiana. *Trans. Am. microsc. Soc.* **20**, 51–5.

Šrámek-Hušek, R. 1957. K poznáni nálevníků ostravského kraje (Zur Kenntnis der Ciliaten des Ostrauer-Gebietes). *Vest. esl. zool. Spol.* **21**(1), 1–24.

Stammer, H. J. 1935. Zwei troglobionte Protozoën; *Spelaeophrya troglocaridis* n.g., n.sp. von den Antennen der Höhlengarnele *Troglocaris schmidti* Dorm. und *Lagenophrys monolistrae* n. sp. von den Kiemen (Pleopoden) der Höhlenasselgattung *Monolistra. Arch. Protistenk.* **84**, 518–27.

Stein, F. 1852 (1851). Neue Beiträge zur Kenntnis der Entwicklungsgeschichte und des feineren Baues der Infusionsthiere. *Z. wiss. Zool.* **3**(1851), 475–509.

Stein, F. 1859a. Characteristik neuer Infusorien-Gattungen. *Lotos* **9**, 2–5.

Stein, F. 1859b. Characteristik neuer Infusorien-Gattungen. *Lotos* **9**, 57–60.

Stein, F. 1860a. Einen Vortrag über bisher unbekannt gebliebene *Leucophrys patula* Ehbg. und über zwei neue Infusoriengattungen *Gyrocorys* und *Lophomonas. Mém. Soc. r. Sci. Bohême* **1860**, 44–50.

Stein, F. 1860b. Eintheilung der holotrichen Infusionsthiere und stellte einige neue Gattungen und Arten aus dieser Ordnung auf. *Mém. Soc. r. Sci. Bohême* **1860**, 56–62.

Stein, F. 1863 (1862). III. Section: Zoologie und vergleichende Anatomie. *Amtl. Ber. Dt. Naturf. u. Aerzte* **37**, 162.

Stein, F. 1867. *Der Organismus der Infusionsthiere nach eingenen Forschungen in Systematischer Reihenfolge bearbeitet.* II Leipzig. 355pp.

Stokes, A. C. 1884. Notes on some apparently undescribed forms of freshwater Infusoria. *Am. J. Sci.* **28**, 38–49.

Stokes, A. C. 1885a. Some new Infusoria from American Fresh Waters. *Ann. Mag. nat. Hist.* (Ser. 5) **15**, 437–49.

Stokes, A. C. 1885b. Some new Infusoria. *Am. Nat.* **19**, 433–43.

Stokes, A. C. 1890. Notices of some new fresh-water Infusoria. *Proc. Am. phil. Soc.* **28**, 74–80.

Stout, J. D. 1960. Morphogenesis in the ciliate *Bresslaua vorax* Kahl and the phylogeny of the Colpodidae. *J. Protozool.* **7**(1), 26–35.

Strand, E. 1928 (1926). Miscellanea nomenclatorica zoologica et palaeontologica. *Arch. Naturgesch.* **92**(A8), 31.

Suhama, M. 1969. The fine structure of ectoplasmic organelles of *Tillina canalifera* (Ciliata, Trichostomatida) from the trophic stage to the division cyst stage. *J. Sci. Hiroshima Univ.* (Ser. B, div 1), **22**, 103–18.

Swarczewsky, B. 1928a. Zur Kenntnis der Baikalprotistenfauna. Die an den Baikalgammariden lebenden Infusorien. I. Dendrosomidae. *Arch. Protistenk.* **61**, 349–78.

Swarczewsky, B. 1928b. Zur Kenntnis der Baikalprotistenfauna. Die an den Baikalgammariden lebenden Infusorien. II. Dendrocometidae. *Arch. Protistenk.* **62**, 41–79.

Swarczewsky, B. 1928c. Zur Kenntnis der Baikalprotistenfauna. Die an den Baikalgammariden lebenden Infusorien. III. Discophryidae. *Arch. Protistenk.* **63**, 1–17.

Swarczewsky, B. 1928d. Zur Kenntnis der Baikalprotistenfauna. Die an den Baikalgammariden lebenden Infusorien. IV. Acinetidae. *Arch. Protistenk.* **63**, 362–409.

Swarczewsky, B. 1928e. Zur Kenntnis der Baikalprotistenfauna. Die an den Baikalgammariden lebenden Infusorien. V. Spirochonina. *Arch. Protistenk.* **64**, 44–60.

Tamar, H. 1971. *Mesodinium fimbriatum* Stokes, 1887, a ciliate with bifurcated and barbed cirri. *Acta Protozool.* **9**(12), 209–22.

Tamar, H. 1973. Observations on *Askenasia volvox* (Claparède & Lachmann, 1859). *J. Protozool.* **20**(1), 46–50.

Tatem, J. G. 1869. On a new Infusorium. *Mon. microsc. J.* **1**, 117–18.

Thompson, J. C. and Corliss, J. C. 1958. A redescription of the holotrichous ciliate *Pseudomicrothorax dubius* with particular attention to its morphogenesis. *J. Protozool.* **5**(3), 175–84.

Tucolesco, J. 1962. Protozoaires des eaux souterraines. 1. 33 Espèces nouvelles d'infusoires des eaux cavernicoles roumaines. *Annls Spéléol.* **17**, 89–105.

Tuffrau, M. 1953. Les processus cytologiques de la conjugation chez *Spirochona gemmipara* Stein. *Bull. biol. Fr. Belg.* **87**, 314–22.

Tuffrau, M. 1967. Perfectionnements et pratique de la technique d'impregnation au protargol des infusoires ciliés. *Protistologica* **3**(1), 91–8.

Turner, J. P. 1937. Studies on the ciliate *Tillina canalifera* n. sp. *Trans. Am. microsc. Soc.* **56**, 447–56.

Vörösváry, B. 1950. A 'Kalános Patak' Csillós Véglényei. *Ann. Biol. Univ. Szeged* **1**, 343–87.

Vuxanovici, A. 1959a. Contribution to the study of the genus *Loxophyllum*. *Revue Biol. Buc.* **4**, 165–74.

Vuxanovici, A. 1959b. Contributii la studiul unor Infuzorii Holotrichi. *Studii Cerc. Biol.* (Biol. Anim.) **11**, 307–35.

Vuxanovici, A. 1960. Contributii la studiul specilor grupei subgenurilor *Lionotus-Hemiophrys* (Ciliata). *Studii Cerc. Biol.* (Biol. Anim.) **12**, 125–39.

Vuxanovici, A. 1962a. Contributii la sistematica ciliatelor (Nota I). *Studii Cerc. Biol.* (Biol. Anim.) **14**, 197–216.

Vuxanovici, A. 1962b. Contributii la sistematica ciliatelor (Nota II). *Studii Cerc. Biol.* (Biol. Anim.) **14**, 331–49.

Vuxanovici, A. 1963. Contributii la sistematica ciliatelor (Nota IV). *Studii Cerc. Biol.* (Biol. Anim.) **15**, 65–93.

Wang, C. C. 1931. On two new ciliates (*Holophrya latercollaris* sp. nov. and *Choanostoma pingi* gen. and sp. nov.). *Contr. biol. Lab. Sci. Soc. China* **6**(10), 105–11.

Weinreb, S. 1955a. *Homalozoon vermiculare* (Stokes): I. Morphology and Reproduction. *J. Protozool.* **2**(2), 59–69.

Weinreb, S. 1955b. *Homalozoon vermiculare* (Stokes): II. Parapharyngeal granules and trichites. *J. Protozool.* **2**(2), 67–70.

Wenrich, D. H. 1929a. Observations on some freshwater ciliates. I. *Teuthophrys trisulca*, Chatton et de Beauchamp, and *Stokesia vernalis* n.g., n.sp. *Trans. Am. microsc. Soc.* **48**, 221–37.

Wenrich, D. H. 1929b. Observations on some fresh-water ciliates (Protozoa) II. *Paradileptus*, n. gen. *Trans. Am. microsc. Soc.* **48**, 352–65.

Wenrich, D. H. 1929c. The structure and behaviour of *Actinobolus vorax. Biol. Bull. mar. biol. lab. Woods Hole* **56**, 390–401.

Wenzel, F. 1953. Die Ciliaten der Moosrasen trockner Standorte. *Arch. Protistenk.* **99**, 70–141.

Wenzel, F. 1955. Über eine Artentstenhung innerhalb der Gattung *Spathidium* (Holotricha, Ciliata). *Arch. Protistenk.* **100**, 515–40.

Wenzel, F. 1969. *Stammeridium* nom. nov. Eine nomenklatorische Korrektur. *Arch. Protistenk.* **111**, 275.

Wessenberg, H. and Antipa, G. 1969 (1968). Studies on *Didinium nasutum* I. Structure and ultrastructure. *Protistologica* **4**(4), 427–48.

Wetzel, A. 1928. Der Faulschlamm und seme Ziliaten Leitformen. *Z. Morph. Okol. Tiere* **13**, 179–328.

Wilbert, N. 1972 (1971). Morphologie und Ökologie einiger neuer Ciliaten (Holotrica, Cyrtophorina) des Aufwuchses. *Protistologica* **7**(3), 357–63.

Wilbert, N. and Schmall, G. 1976. Morphologie und Infraciliature von *Coleps nolandi* Kahl, 1930. *Protistologica* **12**(1), 193–7.

Woodruff, L. L. and Spencer, H. 1922. Studies on *Spathidium spathula* I. The structure and behaviour of *Spathidium*, with special reference to the capture and ingestion of its prey. *J. exp. Zool.* **35**, 189–205.

Wrzesniowski, A. 1870. Beobachtungen über Infusorien aus Umgebung von Warschau. *Z. wiss. Zool.* **20**, 467–511.

Zacharias, O, 1893. Faunistische und biologische Beobachtungen am Gr. Plöner See. *Forsch Ber. biol. Stn Plön* **1**, 1–52.

Zenka, W. 1866. Beiträge zur Naturgeschichte der Infusorien. *Arch. mikrosk. Anat.* **2**, 332–48.

Taxonomic index

Subclasses, orders, suborders and families are in capital letters. For genera the correct names are in *italic*; synonyms are in lower case roman.

385